# Computer Communications and Networks

**Series editor**
A.J. Sammes
Centre for Forensic Computing
Cranfield University, Shrivenham Campus
Swindon, UK

The **Computer Communications and Networks** series is a range of textbooks, monographs and handbooks. It sets out to provide students, researchers, and non-specialists alike with a sure grounding in current knowledge, together with comprehensible access to the latest developments in computer communications and networking.

Emphasis is placed on clear and explanatory styles that support a tutorial approach, so that even the most complex of topics is presented in a lucid and intelligible manner.

More information about this series at http://www.springer.com/series/4198

Jesús Téllez • Sherali Zeadally

# Mobile Payment Systems

## Secure Network Architectures and Protocols

Springer

Jesús Téllez
University of Carabobo
Valencia, Carabobo, Venezuela

Sherali Zeadally
University of Kentucky
Lexington, KY, USA

ISSN 1617-7975 ISSN 2197-8433 (electronic)
Computer Communications and Networks
ISBN 978-3-319-23032-0 ISBN 978-3-319-23033-7 (eBook)
DOI 10.1007/978-3-319-23033-7

Library of Congress Control Number: 2017948868



Printed on acid-free paper

This Springer imprint is published by Springer Nature
The registered company is Springer International Publishing AG
The registered company address is: Gewerbestrasse 11, 6330 Cham, Switzerland

*To the most important persons in my life, my wife, Raiza, and my daughter, Alexandra, whose sacrifice was essential to allow me to finish this work. I also dedicate this book to my parents who are no longer physically in this world but will live in my mind and heart until my last breath. Finally, but no less important, I cannot forget my friends, brothers and sisters, students, and coworkers who have supported me in my professional career.*

Jesús Téllez

*To my wife, Borrara; my daughters, Zobia and Zofia; and my parents.*

Sherali Zeadally

# Preface

In the last two decades, we have seen significant technological advances in computing, networking, storage, and processing technologies. All these advances have led to the emergence of various types of networks, all kinds of user applications, and a wide range of computing devices whose costs keep decreasing while their performance keeps improving. These new capabilities have paved the way for users to have the ability to access information anywhere, anytime from any personal computing device. Additionally, we have also witnessed the rapid development of mobile networking technologies and communication protocols that have accelerated the design and deployment of mobile payment systems (MPSs) which enable payments for services and goods from mobile devices using different methods, such as credit card payments, micro-payments, and digital coins. Security issues in these payment systems have become increasingly important for all parties involved (customers, merchants, banks, etc.) during their interactions. Security is one of the most important aspects that must be taken into consideration when designing, implementing, and deploying secure payment systems. It is essential to mitigate threats, vulnerabilities, and risks that affect such systems. While supporting secure payment transactions, high performance is also another major objective that must be met by any MPS even in the presence of constraints of portable devices, limited communication capabilities, and other limitations.

The main aim of this book is to present state-of-the-art research achievements and results in the field of MPSs. We expect this book to be a valuable and an authoritative reference for designers, developers, engineers, students, faculty members, and researchers who have a strong interest in mobile commerce and in particular mobile payment systems. We have organized this book into five chapters: "(1) Introduction"; (2) "Mobile Device Security"; (3) "Architectures and Models for Mobile Payment Systems"; (4) "Security in Mobile Payment Systems"; and (5) "Future Challenges and Opportunities."

*Mobile Payment Systems: Secure Network Architectures and Protocols* is an in-depth literature review of recent research results obtained by experts and researchers around the world, along with the authors' results in the area of MPSs. We present here a summary of the chapters in this book.

Chapter 1, "Introduction", presents the basic and fundamental concepts related to MPSs. In this chapter, we give an overview of mobile commerce, mobile payment characteristics, existing mobile payment methods, mobile payment stakeholders, and technologies for mobile payments. The chapter further discusses the benefits and disadvantages of MPSs and the general entities that underpin such systems.

In Chap. 2, "Mobile Device Security", we review the mobile devices currently available on the market that can perform a variety of tasks. We also discuss the operating systems (OSs) available on the market based on the requirements of mobile devices and the advantages and disadvantages of each OS. We further highlight some important differences between mobile device security and personal computer (PC) security to improve our understanding of mobile device security. The chapter also presents a threat model, which identifies the threats against MPSs, the resources to be protected, the features of the attackers, and the potential attack vectors. Finally, it highlights two main mechanisms that are used to avoid various kinds of threats for smartphones: intrusion detection systems and trusted mobile-based solutions.

In Chap. 3, "Architectures and Models for Mobile Payment Systems", we review the mobile payment models that have been proposed in the literature in the last decade. We classify these models based on the features and technology mostly used by end users. The main goal of this classification is to help readers stay up to date with the state-of-the-art mobile payment models proposed in the literature based on their core features such as micro-payments, cryptographic technique (asymmetric cryptographic and symmetric cryptographic), technology used (short message service, biometric technology, radio-frequency identification (RFID) technology, near-field communication (NFC), 2-D barcode technology, and peer-to-peer technology), Session Initiation Protocol (SIP), communication restriction, mobile agent technology, and wireless application protocol.

Chapter 4, "Security in Mobile Payment Systems", focuses on transaction security properties that must be satisfied by MPSs to ensure that all the information exchanged between the mobile payment user and the entity with which he/she is communicating is secure at all times during the payment transaction made using mobile devices. We present a brief review of the most commonly used cryptography schemes for secure communications among the parties involved. This chapter also presents a summary of the types of vulnerabilities and threats and their corresponding risks in an MPS environment together with relevant protection solutions. Furthermore, we discuss some obstacles that make the design of secure MPSs still a major challenge.

In Chap. 5, "Future Challenges and Opportunities", we outline some of the challenges posed by MPSs which rely on new emerging technologies that offer attractive business opportunities. This chapter also discusses security challenges that must be addressed in order to establish a secure environment for ubiquitous mobile commerce (m-commerce). We also identify some of the challenges faced by users in MPSs due to their lack of payment options around their mental model development, usability issues, prepurchase anxiety, trust issues, ease-of-use, and support for routine purchases with mobile payments. We also discuss alternate cryptographic schemes such as elliptic curve cryptography and self-certified public keys and their applications to MPSs. This chapter also highlights opportunities offered by mobile cloud computing (MCC) and vehicular ad hoc networks (VANETs) in the design and development of MPSs.

We thank Simon Rees and Wayne Wheeler at the Springer office in London, United Kingdom, for their constant encouragement, support, and advice throughout the preparation of this book.

Valencia, Venezuela — Jesús Téllez
Lexington, USA — Sherali Zeadally

# Contents

# Chapter 1
# Introduction

## 1.1 Electronic Money

Since ancient times, man has devised various schemes to give value to things and to be able to trade them. The first method man used to get the things he wanted or needed was called *Barter* (see Fig. 1.1). This was efficient while human needs were few and limited. But as societies became more complex and commercial activity increased, barter became insufficient, and man had to adopt other systems as units of exchange change and measures of value. Thus emerged the concept of money, which has been one of the most important inventions in the history of humanity [19].

According to Kelley [83], money has three social functions: (a) as a unit of account, or a way to measure and record value (a pig is worth X dollars); (b) as a way to store value conveniently for future use (possession of a pig is replaced by possession of a bank account recorded in, and retrievable in, money); and (c) as medium of exchange (instead of having first to find a pig to trade for cloth, money buys cloth or a pig). These functions can only be performed if the money satisfies the following requisites: it must be easily and widely recognizable and difficult to forge (counterfeit), have a fairly value, be suitable and inexpensive for use in daily commerce, and it cannot be manipulated to change its value. There are many payment instruments (paper, credit and debit cards, checks, and coins) that currently satisfy these requirements and are widely accepted and considered to be traditional payment methods, which leads one to wonder if there is a real need to create new forms of money.

With the passage of time, the concept and use of money has evolved to major degrees of abstraction. Though they offer several definitions for money, economists traditionally have defined it as a means of socially accepted exchange, an abstract representation of a value (independent from the inherent value in the paper or metal) endorsed by an authority and generally accepted for the accomplishment of commercial exchanges [30].



J. Téllez, S. Zeadally, *Mobile Payment Systems*, Computer Communications and Networks, DOI 10.1007/978-3-319-23033-7_1

**Fig. 1.1** A man offering chickens in exchange for his yearly newspaper subscription [120]

The increasing development of information and communication technologies (ICTs) allowed the world- wide propagation of the Internet and the emergence of a business environment, *electronic commerce* or *e-commerce*, whose physical location the buyer cannot easily identify and where the realized transactions are done electronically and can be observed by other entities foreign to the operation [30, 192].

In this new type of commerce, payment transactions are based on *remote interaction* where there is no possibility of direct contact between the payer and the payee. Therefore, traditional payment instruments cannot be used in electronic commerce because they are intended for use in payment transactions carried out through a *face-to-face interaction*, where the payer and the payee have the possibility of direct contact.

As stated in literature, the aforementioned situation is the main reason for the emergence of *electronic money*, which refers to all types of stored value cards (also called prepaid cards) and other means to create new forms of money that represent an alternative to the instruments issued or guaranteed by the government. Electronic money is also intended to behave similarly to traditional money but by replacing the traditional support with electronic support, or, in other words, changing the paper to bits.

### 1.1.1 Advantages and Disadvantages of Electronic Money

Undoubtedly, all technology involves a benefit to the user but also creates certain risks and disadvantages. As electronic money does not escape this reality, some advantages and disadvantages of this type of money are shown in the following paragraphs.

The use of electronic money as a payment model has the following advantages [30]:

- It maintains the anonymity of the client.
- It does not require that the merchant be connected with the bank during a transaction.
- The transaction can be realized without the presence of the client.
- It is portable and easy to use.

Nevertheless, despite the advantages discussed above, electronic money presents the following disadvantages [30, 155]:

- It requires specialized equipment for use.
- The processing time and cost of the transactions both increase.
- Overuse of the electronic coins.
- It is susceptible to adulteration.
- It is not easy to follow, therefore money laundering problems are created.

### 1.1.2 Characteristics of Electronic Money

An electronic money system can be considered as such if it has certain properties. The number of properties that the system fulfills determines its quality. A set of desirable properties of electronic money systems are as follows [133]:

- *Independence*: Since the electronic money must be sent through the network, its security cannot depend on the security of the network.
- *Security*: Because electronic currency consists of bits that can be digitally copied and reused easily, mechanisms that allow for the establishment of the authenticity and reuse of these currencies are necessary.
- *Privacy*: The anonymity of the buyer should be guaranteed in valid transactions. This means that when a buyer uses electronic coins his/her identity should not be identifiable through their purchases. However, if some kind of fraud is attempted, the bank is able to identify the buyer in order to take necessary legal actions.
- *Off-line Payment*: Transactions must be performed online. When a transaction is performed between a buyer and a merchant, the merchant should be connected with the bank to check the buyer's payment.

- *Transferability*: The money can be transferred to other users who may use those electronic coins later without any problems.
- *Dividability*: A "piece" of money $C$ can be divided into other smaller denominations. The sum of the value of the subdivided pieces must equal $C$.

## 1.2 Electronic Commerce

The face-to-face exchange of goods between two parties dates back to before the beginning of recorded history. The worldwide growth of the Internet has allowed the birth of an extensive spectrum of services (such as file transfer, electronic mail, etc.) where the most popular is e-commerce, a business environment which enables electronic transfer of transactional information [9, 192].

Electronic commerce is a rising technological wave that has flourished because of the openness, speed, anonymity, digitization, and global accessibility characteristics of the Internet, which has facilitated real-time business activities, including advertising, querying, sourcing, negotiating, auctioning, ordering, and paying for merchandise [92, 192].

In the literature, there exists more than a definition for e-commerce but, in general terms, it includes any normal business function performed via electronic networks. However, in this work, we follow the definition given by [127]:

- From a *communications perspective*, electronic commerce refers to the process of delivering information, products, services, or payments through different means such as telephone lines, computer networks, etc.
- From a *business process perspective*, electronic commerce refers to the usage of technology for automation business transactions and workflow.
- From a *service perspective*, electronic commerce is considered a tool for companies, consumers, and management to reduce the cost of services while enhancing the quality of goods, and increasing the speed of service delivery.
- From an *online perspective*, electronic commerce offers the capacity to buy and sell products, information, and other online services via the Internet.

E-commerce embraces all the features of commerce (such as catalogues, online purchases, customer service, products specification, etc.), and creates new features in the context of this type of commerce. Concepts that have emerged in electronic commerce include *e-payment*, *e-banking*, *e-lottery*, and *e-government*, involving actions such as signing contracts, transferring funds, and distributing intangible digital goods [54].

Electronic commerce allows people to conduct business and communicate via the Internet, simulating and improving the traditional forms. For example, electronic mail (or e-mail) improves the time of delivery for communication. Using the traditional route, a person would have to wait days to receive letters; in contrast, via

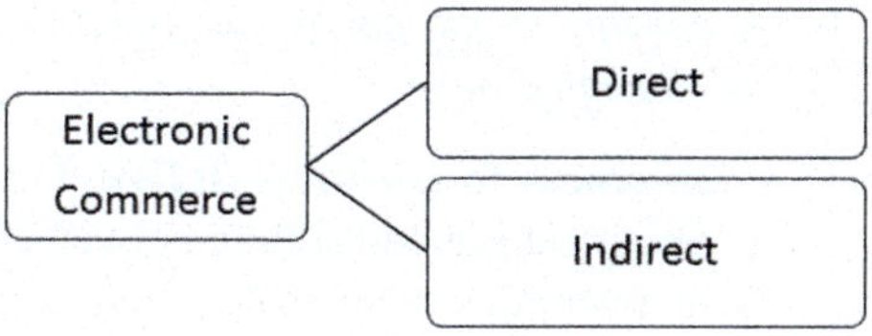


**Fig. 1.2** Classification of the electronic commerce according to the goods or services that are commercialized

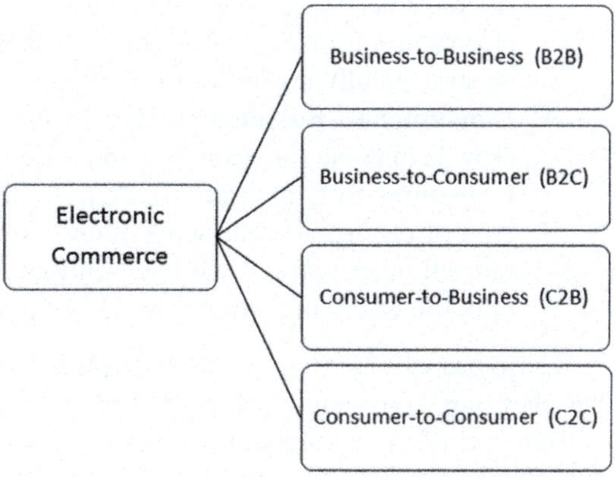


**Fig. 1.3** Classification of the electronic commerce according to the parties engaged in the transaction

the electronic route, this process is realized in minutes, thereby reducing the time of delivery and the cost of sending without losing the characteristics of paper mail.

Regarding business transactions, there are diverse e-commerce websites that permit customers to examine the goods and services offered in the online stores (which offer both electronic and tangible products), to select the products or services they desire, and to pay for them using both credit or debit cards 24 h a day, 7 days a week from their PCs or mobile devices.

Figure 1.2 shows the classification of electronic commerce regarding the goods or services sold, while Fig. 1.3 shows the classification of electronic commerce regarding the parties engaged in the transaction. The details about this classification of electronic commerce follow.

(a) *According to the goods or services that are commercialized, electronic commerce can be classified as follows:*

- **Electronic Commerce Direct:** E-commerce in which the order, the payment, and the sending of the intangible goods and/or services take place online. The transactions or operations linked with banking services and the sale of tickets (theatre, concert, etc.), trips and software are some examples of this category.
- **Electronic Commerce Indirect:** E-commerce consisting of the acquisition of tangible goods (e.g., books, flowers, perfumes, etc.) that must be sent physically, using channels or traditional routes of distribution.

(b) *According to the parties engaged in the transaction, electronic commerce can be classified into:*

- **Business-to-Business (B2B):** It refers to the exchange of digital information about buying and selling between businesses, e.g., suppliers of raw materials and supplies, wholesalers, retailers, distributors, and others.
- **Business-to-Consumer (B2C):** It occurs when a company sells its products to consumers (e.g., www.bn.com, www.amazon.com, etc.). In this type of electronic commerce, the company can gather diverse information that could be strategically useful for the company.
- **Consumer-to-Business (C2B):** In this category, the consumers gather in crowds in cooperatives to bid and acquire products at a better price.
- **Consumer-to-Consumer (C2C):** Better known as Internet auctions, this type of electronic commerce allows consumers to offer goods and services to each other without the intervention of a company. Examples of this type of commerce can be found in www.deremate.com and www.ebay.com.

The types of electronic commerce presented in Fig. 1.3 are not the only ones, but they are considered the principal ones that companies offer worldwide to consumers and other companies with the aim of reducing operative costs, improving communication, and allowing for the completion of transactions online regardless of the geographical location of any party involved.

## 1.3 Mobile Commerce

The rapid development of mobile network technology and advances in portable devices have allowed people to access the Internet using mobile devices anywhere and anytime in order to perform e-commerce transactions and another operations such as web browsing, email reading, and so on.

The wireless revolution revealed that mobile devices are becoming a critical element of the new digital economy in which transactions are experiencing a rapid transition from fixed locations to anytime, anywhere, and anyone [7].

*Mobile e-commerce*, or simply *m-commerce*, alludes to electronic commerce transactions performed through wireless networks using mobile devices. It represents an alternative to e-commerce or a normal extension of electronic commerce and a channel for commerce [162].

In recent years, m-commerce has received considerable attention because it has grown fast. A novel survey from ABI Research (a marketplace firm that specializes in global connectivity and the nascent technology) reviewed by ZippyCart Ecommerce Software Content Team forecasts that, by the end of 2015, global revenues of m-commerce will reach approximately US$ 119 billion [201].

M-commerce, as an expansion of e-commerce, has both characteristics common to e-commerce and other unique qualities that provide customers with additional benefits and added value. Some advantages that mobile electronic commerce provides in comparison to electronic commerce that is performed in fixed networks are as follows [157]:

- MPSs allow users to connect to the Internet in order to execute e-commerce transactions using their mobile devices in real time at any place.
- Mobile devices are lighter and smaller than PCs, which allows them to be taken anywhere easily.
- Mobile devices can be customized to suit the requirements of particular users since these devices are personal. This allows a mobile operator to provide specialized services to users based on their characteristics and location.

## 1.4 Mobile Payment

One of the key requirements of transactions in e-commerce is a method to pay for any item purchased. This type of payment, which is made electronically, is called electronic payment or *e-payment* and refers to the exchange of monetary values across a network of computers [138, 167, 182]. E-payment is considered as an alternative to those payments which are generally made physically in the merchant's premises using a credit or debit card and accompanied by a receipt as proof of the transaction.

Despite the existence of various types of electronic payment methods, the bank transfer represents the most simple payment method in which an e-commerce website must send, via an e-mail, an invoice to the customer with the total amount payable for the goods and/or services purchased and the bank account number where the money should be transferred to complete the payment. Upon receipt of payment, the website must send the corresponding payment receipt to the client via email before delivering to the customer the goods and/or services purchased [92].

The scenario described here demonstrates that e-commerce websites that use bank transfers as their payment method may allow a customer to make purchases *online* but will require payment *off-line*. Therefore, many e-commerce websites nowadays are not using this traditional electronic payment method but instead are using new payment systems that allow transactions to be completed totally online using credit cards, electronic checks, electronic cash, and so on.

Payment by credit card is the most popular type of e-payment. Goods must be selected from an online store before a customer can perform a payment through the credit card supporting the platform provided by the store. To authenticate a customer and obtain authorization for the payment, a form must be completed with relevant customer information such as billing address, credit card number, and date of birth. Then, this information must be sent to the customer's credit-card issuer to verify that the credit is available and to approve or deny the payment. If the request is approved, the customer will receive the goods and the related payment receipt.

Most of the credit card payment protocols employ a 128-bit SSL (Secure Socket Layer) protocol [41]) to secure communications via the Internet. However, these payment schemes based on SSL provide less security for payment transactions than other credit card payment systems like SET (Secure Electronic Transaction) [114]. In payment systems based on SET, customers need to install the SET payment software (known as the SET wallet) on their computers. The software is activated after the customer has browsed the goods and/or services from a online store that supports SET. Then, the necessary information for the payment is completed and the SET wallet generates a purchase request (by performing extremely secure cryptographic operations such as public-key cryptographic) that is sent to both the store and the customer's credit card company for payment authorization. If the payment is approved, the store's account receives the requested amount that has transferred from the customer's account before delivering the purchased goods and the corresponding payment receipt [27, 92].

Credit cards are frequently used as a basic payment method for goods and/or services on the Internet. This is due to their popularity and because people also use them to buy goods and/or services in physical stores. Nevertheless, the operational cost of a credit card payment system is high, particularly for businesses, making it an unsuitable method for low-value transactions.

*Micro-payment*, a payment method suitable for low-value transactions, reduces the operational costs by deploying light computational cryptographic operations and simple message passing. Some examples of these kinds of systems include Chipper, GeldKarte, Mondex, Proton, NetBill, KLELine, Odysseo, MicroMint, Floodgate, and Payword [31, 123, 141, 149].

*M-payment* is a new concept introduced by e-payment in wireless environments and refers to any payment transaction that implies buying of goods or services, executed via a device with wireless capability [167]. Thus, purchases can be made by the clients in online stores, which are able to accept payments executed from mobile devices whilst on the move [90].

As an important aspect of electronic commerce for fund transfer between the entities of a payment system that have agreed to purchase or sell products and/or services, electronic payments (including mobile e-payment) have to be realized in a safe way to reduce concerns of customers who make online payments via Internet stores. Therefore, it is important to incorporate mechanisms in personal mobile devices that will securely authenticate transaction requests and that can be used for (preferably) multiple transactions and scenarios. The security of transactions is one of the biggest challenges that mobile payment systems must overcome [65, 199].

### *1.4.1 Mobile Payment Characteristics*

For Carr [21], a mobile payment service can become acceptable in the market as a mode of payment by satisfying the following conditions:

I. **Simplicity and Usability:** The learning curve of the m-payment application must be small or none for the customer. Also, the application should be friendly for the user and customizable for his/her convenience.

II. **Universality:** M-payment services must support transactions between one customer and another customer (C2C), from a business to a customer (B2C), or between businesses (B2B). The coverage should include domestic, regional, and global environments. Payments must be possible in terms of both low-value micro-payments and high-value macro-payments.
III. **Interoperability:** Open technologies and standards must be used to develop MPSs capable of interacting with other systems once they have been implemented.
IV. **Security, Privacy and Trust:** Mobile payment application providers must be trustworthy for customers in that they will not misuse customers' credit or debit card information. Secondly, customer privacy must be preserved when these transactions are registered in order to allow that the credit histories and consumption patterns of the customer are usable for public investigation. Mobile payments must be anonymous in the same way as cash transactions are. Third, the system should be foolproof and resistant to attacks from hackers and cyberterrorists. This may be achieved by the use of public key infrastructure security, biometrics, and passwords integrated into the mobile payment solution architectures.
V. **Cost:** To the extent possible, m-payments must be less expensive in comparison with existing payment mechanisms. Moreover, with respect to other modes of payment, an m-payment solution should compete in terms of cost and convenience.
VI. **Speed:** Customers and merchants expect that m-payments will be executed at an acceptable velocity.
VII. **Cross Border Payments:** The m-payment application should be accessible globally in order to be widely accepted.

## *1.4.2 Existing Mobile Payment Methods*

Existing systems are classified and compared according to several standards (see Fig. 1.4).

### 1.4.2.1 Basis of Payment

Current mobile payment methods in use or under test can be classified by taking into consideration the way that money transfer is structured [56, 92, 199]:

- *Account-based:* Every consumer is associated with a specific account that an Internet Payment Provider (IPP) maintains. Debits and credits are exchanged during a transaction. The consumer is then periodically billed and pays the balance of the account to the IPP [56, 199].

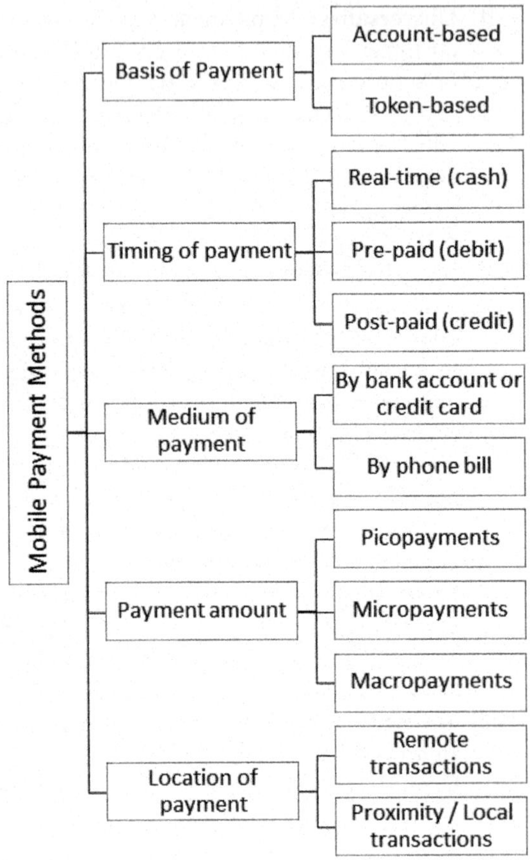


**Fig. 1.4** Classification of existing mobile payment methods

Due to the high charges, traditional account-based payment methods are generally not suitable for handling transactions of very low value (i.e., micro-payments).

- *Token-based:* Using of electronic tokens is the alternative to maintaining an account for each consumer. A token, which is used during a transaction, is a medium of exchange representing some monetary value usually supported by a bank. Before beginning a transaction, it is necessary that the consumer transform his/her actual currency into electronic format (i.e., tokens) by soliciting the issuer to subtract the solicited amount from his/her account. Each merchant sends to his/her acquirer all the tokens collected to redeem the money that will be transferred to the merchant's account [56, 199].

  In comparison with account-based payment systems, the token-based payment method offers lower operational cost since the client does not need payment authorization from the bank for each transaction. Therefore, these types of payment methods are more suitable for handling micro-payments [92].

#### 1.4.2.2 Timing of Payment

Mobile payment can be performed at different times [27, 56, 199]:

- *Real-time (cash):* Payment methods choose the real-time or cash-like payment schedule that involves some type of electronic currency that is exchanged at the time of a transaction [56, 199]. Examples of these types of payment methods include beenz.com[1] and eCash [156].
- *Pre-paid (debit):* Consumers must pay before they receive the product desired. Thus, the payer's local/central account should be credited with an initial amount of money before buying. Then, with successive purchases, the balance of the account decreases. A variant of the pre-paid payment, called pay-now, occurs when there exists an online connection between the payee and the issuer (who holds the central account of the payer).

  Smart cards[2] that store value and electronic "wallets" (e.g., reloadable Visa cash cards, Proton-based cards) are examples of pre-paid payment methods [27, 56, 199].
- *Post-paid (credit):* The content is received and consumed by consumers before payment is made. To achieve this, the consumers must be authenticated by IPP to verify if he/she has enough funds for the purchase. This means that the balance of the payer's local/central account is credited with an initial amount of money before purchases are made. Note that the payer's balance initially becomes negative. Electronic checks and credit cards are examples of credit-based payment methods [27, 56, 199].

#### 1.4.2.3 Medium of Payment

Mobile payment methods can be classified according to the medium of payment as follows [56, 199]:

- *Mobile payment by bank account or credit card:* In this type of payment, the following categories can be identified:
  - When the payment is executed, the card is not accessed in a direct manner, which means that the payment systems are built taking into consideration the conventional payment methods such as direct debit or credit card. Paybox [92] is an example of this kind of mobile payment systems.

[1]Beenz.com was a web site founded in 1998 that allowed consumers to earn beenz, a type of online currency, for performing activities such as visiting a website, shopping online, or logging on through an Internet service provider. The beenz e-currency could then be spent with participating online merchants.

[2]Smart-cards were introduced in wireless networks as SIM cards. They provided superior fraud protection since they were specifically designed to secure data against physical and logical attacks [2].

  - When the payment is executed, the card is accessed in a direct manner, which means that the payment card is placed into the mobile phone and can be read directly by the phone. Phones with two slots and two chips are examples of this kind of solution.

- *Mobile payment by phone bill:* In this method, settlement is made via phone bill for common mobile payment. An example is NTT DoCoMo's i-mode, which uses its portal to provide this type of payment method [132]. Mobile payments via phone bill have an advantage since banks and credit card companies are excluded to some degree to simplify the process, which implies a cost reduction but requires a reliability of the mobile operators' bills.

#### 1.4.2.4 Payment Amount

Depending on the quantity of money being transfered from the customer to the merchant, mobile payment methods can be classified as follows [116]:

- *Picopayments*, for those transactions whose amount is less than US$0.10.
- *Micro-payments*, for amounts between US$0.11 and US$10.
- *Macro-payments*, for amounts exceeding US$10.

#### 1.4.2.5 Location of Payment

Payments supported on location can be classified as follows [146]:

- *Remote transactions*: Refers to those transactions that can be performed independently of the users' location. Examples include mTickets, digital cash, peer-to-peer payments, and so on.
- *Proximity/local transactions*: Refers to those transactions in which the mobile device communicates locally to pay for merchandise and services by using short-range messaging protocols such as Bluetooth, infrared, RFID, and contactless chips.

### *1.4.3 Mobile Payment Stakeholders*

Singh et al. [164] recognizes, for the implementation process of mobile payments, various stakeholders including consumers, merchants, mobile network operators, mobile device manufacturers, financial institutions and banks, software and technology providers, and governments. Every player has different motivations and strategies, which can be in conflict; for example, the goal of the customers and the merchants will be the reduction of costs for every m-payment transaction whereas

**Table 1.1** Expectations of each of the stakeholders in mobile payment [21]

| Stakeholder | Expectations |
|---|---|
| Consumer | • Personalized service<br>• Minimum learning curve<br>• Trust, privacy, and security<br>• Ubiquitous use anywhere, anytime, and any currency<br>• Low or zero cost of usage<br>• Interoperability between different network operators, banks, and devices<br>• Anonymity of payments (like cash)<br>• Person-to-person transfers |
| Merchant | • Faster transaction times<br>• Low or zero cost when using the system<br>• Integration with existing payment systems<br>• High security<br>• Customizable services<br>• Real-time status of the mobile payment service |
| Banks | • Network operator independent solutions<br>• Payment applications designed by the bank<br>• Exceptional branding opportunities for banks<br>• Better volumes in banking: more card payments and less cash transactions<br>• Customer loyalty |
| Telecom Network Providers | • New income generation through increased traffic<br>• Increased attractiveness as partner for content providers |
| Mobile Device Manufacturer | • Large market adoption with embedded mobile payment application<br>• Low time to market<br>• Increase in Average Revenue Per User (ARPU) |
| Government | • Revenue through taxation of m-payments |

the intention of the telecommunications network provider will be to increase profits through every m-payment transaction. Table 1.1 describes, for each stakeholder, their expectations.

### *1.4.4 Technologies for Mobile Payments*

Because, basically, a GSM mobile phone utilizes diverse potential channels to send or receive information that impacts how m-payment schemes are implemented, in this section we briefly describe various technological possibilities for implementing m-payments:

- **Short Message Service (SMS):** is a service that allows mobile systems and other network-connected devices to exchange short text messages with a maximum length of 160 characters (due to limitations caused by the way that SMS is transmitted) [86]. Originally designed as part of the Global System for Mobile Communications (GSM), SMS is now available on a wide range of networks.

Although SMS was originally meant to notify users of their voicemail messages, it has now become a popular means of communication by individuals and businesses [108]. For example, banks worldwide are using SMS to provide information about the status of people's accounts (informational) or to transmit payment instructions from the phone (transactional).

- **Unstructured Supplementary Services Delivery (USSD):** USDD is a technology unique to GSM that refers to a capability built into the GSM standard to support the transmission of information over the signaling channels of the GSM network. USSD provides session-based communication, enabling a variety of applications. USSD is session-oriented, transaction-oriented technology, while SMS is a store-and-forward technology. Turnaround response times for interactive applications are shorter for USSD than SMS **Unlike** the asynchronous SMS service, a USSD request opens a session that may induce other network operations or a USSD response before releasing the connection. Mobile-originated USSD may be thought of as a trigger for a network operation. USSD works with any mobile phone since the coded commands are entered in the same way as a phone number (e.g., *123#1234567890#) [158].
- **WAP/GPRS:** WAP/GPRS is a mobile data service available to GSM users that provides packet-switched data for GSM networks. GPRS enables services such as Wireless Application Protocol (WAP) access,[3] Multimedia Messaging Service (MMS),[4] and Internet communication services such as email and world wide web access on mobile phones.
- **SIM-based Application:** The subscriber identity module (SIM) used in GSM mobile phones is a smartcard. Cryptographic algorithms and keys can protect the information in the SIM that makes SIM applications comparatively much more reliable than those applications that stay on the mobile phone. If the SIM card is moved to a new phone, the application will work on the new handset [153].
- **Near Field Communication (NFC):** NFC is a short-range wireless communication standard defined in the ISO/IEC 18092 standard. It is the result of the fusion of a contactless smartcard (RFID) and a mobile phone. NFC communication occurs between two devices (either two active devices or an active device and a passive device) that are within a few centimeters using 13.56 MHz. Active devices are powered by a battery; passive devices gain their energy from the electromagnetic field of the active device [36].

  NFC technology brings the user experience, convenience, and security of contactless technology to mobile devices and is enabling quick transactions and service access in the day-to-day lives of users [40]. Moreover, NFC provides mobile phones with an interface that allows them to act as smartcard readers or to emulate smartcards [36, 136].

[3]WAP allows a mobile phone to surf the Internet. It represents the wireless equivalent of TCP/IP with the benefit of being independent of the bearer [163].

[4]MMS is a standard way to send messages that include multimedia content to and from mobile phones.

- **SIM Application Toolkit (SAT):** SAT is a technology that permits configuring and programming the SIM card. The SIM card contains simple application logic that is able to exchange data with the SMSC to carry out m-payment transactions. The specific mobile operator provides the application logic and is responsible for providing the SIM card. However, the security depends on whether the security mechanisms of the application are implemented or not [85].
- **Voice-based Payment Transaction:** In this type of transaction, payment transactions occur by calling a special telephone number and providing a credit card number. Voice recognition techniques can be used, but they also contain security loopholes [85].
- **I-mode:** In this technology, the content can be placed on the content server without the domain of the gateway because the user has an i-mode subscription; the telephone number is identified by the caller ID and linked directly to a bank account [85, 132].
- **Dual Chip:** Generally, telecom companies buy SIM cards in bulk to customize them for use before sale. Any mobile payment provider that wants to write a mobile payment application in the SIM card, must do so in cooperation with the telecommunications operator (the owner of the SIM). To overcome this limitation, dual chip phones have two slots: one for a SIM card (telephony) and another for a payment chip card.
- **Mobile Wallet:** This technology is an m-payment application software that resides on the mobile phone and contains details of the customer (including his/her bank account details or credit card information) that allow the customer to make payments using the mobile phone. Customers can keep several debit or credit payment instruments in a single wallet. Several implementations of wallets that are company-specific are in use globally.

## 1.5 Mobile Payment System (MPS)

Put simply, an MPS can be defined as a way for consumers to pay for goods and/or services and for merchants to complete transactions using a mobile device in a secure way. However, the literature offers a more complete definition: an MPS is *any conventional or new payment system which enables financial transactions to be made securely from one organization or individual to another over a mobile network (using a mobile device)* [57, 64]. A common gateway and settlement agency is required to provide routing between the various mobile payment providers and for the settlement amongst themselves [164].

There are two basic types of mobile payment devices: (a) one that uses a mobile network to perform the transaction, and (b) one where mobile phones can be used to "talk" with payment receivers in lieu of credit card transactions.

**Some benefits of an MPS are [4]:**

- On-the-go transaction capabilities.
- A wider reach.

- Transaction cost reductions.
- Very large market penetration. There are approximately 4.6 billion people with mobile phones globally. This means that the number of potential customers exceeds that for any other device used today.
- Competitive pricing.
- Order time reduction.
- High anticipation and demand for new technologies and new products.

**Some disadvantages of an MPS are [4]:**

- Less functionality for mobile Internet access on mobile devices compared to wired solutions.
- Limited bandwidth.
- Mobile devices' technology constraints (memory, processing power, display capabilities, input methods).
- Limited battery life.
- Interrupted connectivity.
- Security of data moved across some mobile and wireless networks.
- Potential risks for businesses' investments in hardware and infrastructure as rapid evolution of mobile and wireless technologies continues.
- Limited data throughput speeds compared to wired solutions.
- High risk of loss of or damage to mobile devices.

## *1.5.1 Entities*

Generally, an MPS is made up of the following main engaging parties [78]:

- **Client (C):** A user who buys goods or services from the merchant. This is an entity on the user side equipped with a mobile device with low computational capabilities (e.g., mobile phone, PDA, smartphone, etc.) that has a built-in display, keyboard, and short-range link (such as Infrared, Wi-Fi, Bluetooth, NFC, or RFID). Also, the mobile device has the capability of connecting to the Internet (optional).

  In the context of Vehicular Ad hoc NETworks (VANETs), the client is equipped with an On-Board Unit (OBU)[5] and/or an Application Unit (AU).[6] An AU could be an integrated part of a vehicle that is permanently connected to an OBU or could be a portable device such as a Personal Digital Assistant (PDA), a mobile phone, or a gaming device that can dynamically attach to and detach from an OBU.

[5]An *On-Board Unit (OBU)* is a communication unit equipped with a (short-range) wireless communication device designed for road safety and potentially compatible with other optional communication devices.

[6]An *Application Unit (AU)* is typically a dedicated device that executes a single or a set of applications and utilizes the OBU's communication capabilities.

- **Merchant (M):** An entity that has products or services to sell. It could be a computational one (like a normal web server) or a physical one.
- **Acquirer (A):** The merchant's financial institution responsible for the management of merchant's account, and the verification of the deposited payment instrument.
- **Issuer (I):** The client's financial institution that manages the client's account and afford the electronic payment instruments to be used by client.
- **Payment Gateway (PG):** An additional entity that acts as an intermediary between the acquirer/issuer on the bank's private network side and the client/merchant on the Internet side for payment clearing purposes.

The interactions of the above entities during mobile payment transactions are illustrated in Fig. 1.5 [76].

The client can communicate with the merchant via a short-range link or the Internet by using communication technologies (wired, wireless, or cellular) offered

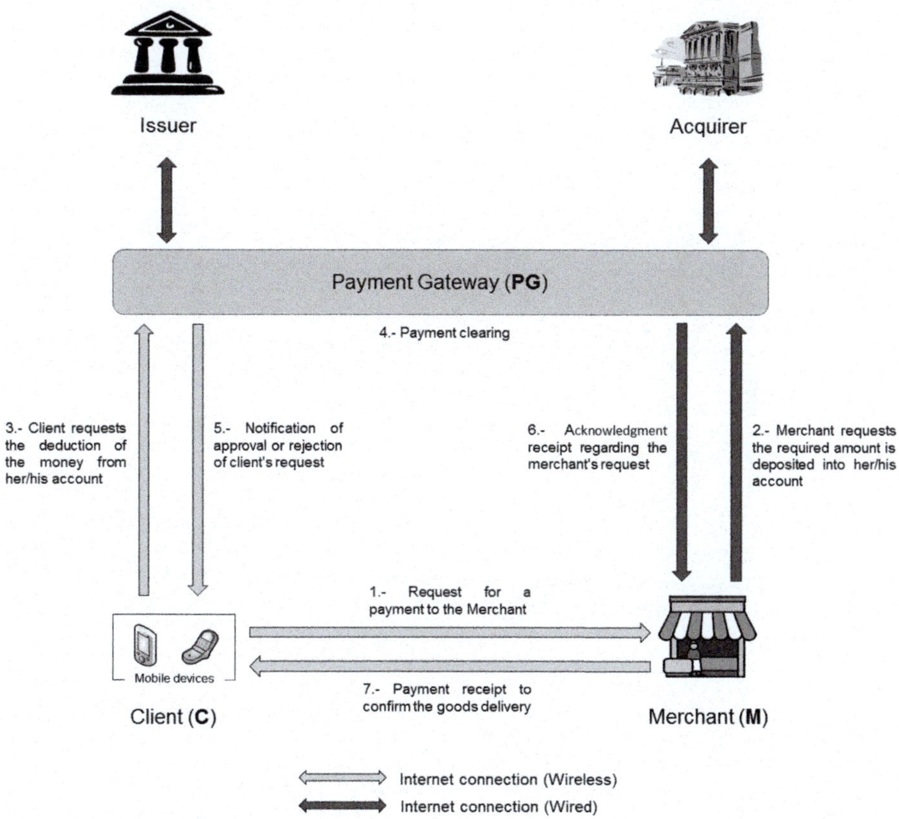


**Fig. 1.5** Traditional entities that make up an MPS and their interactions during a mobile payment transaction

by a mobile phone operator. On the other hand, the connection between the client and the payment gateway or the merchant and the payment gateway can be established through any Internet access service offered by an Internet Service Provider (ISP) or the Internet, also using communication technologies (wired, wireless, or cellular) provided by a mobile phone operator. The connections between the issuer, the acquirer, and the payment gateway typically take place over the private networks of banks (wired), and the communication is made secure using well-known security protocols such as Secure Sockets Layer (SSL) and Transport Layer Security (TLS) [76].

# Chapter 2
# Mobile Device Security

## 2.1 Mobile Devices

A mobile device, which is also referred to as a cell phone device, handheld, handheld device, or handheld computer, is typically a small computing device that can perform a variety of tasks. Also, frequently they are called pocket-sized computers because of the way they are operated and transported.

The first mobile devices started to appear in the late 1970s. The early mobile phones have very little in common with current mobile devices, except the ability to make phone calls. A similar comparison is evident between early personal organizers and more recent PDAs. The only thing they have in common is the ability to hold an address book or a calendar. Therefore, it is possible to state that original mobile devices do not resemble the current generation of mobile devices because they were designed for one particular application or task, while present mobile devices are engineered to be able to execute a variety of tasks. To achieve the desired versatility, mobile devices come with operating systems that allow people to install additional programs. Also, these devices are equipped with features like network connectivity, email, and so on, which make them just as powerful as a computer.

### *2.1.1 Classification of Mobile Devices*

Nowadays, the wide variety of mobile devices available in the market allows users to solve different problems and meet different needs. The devices range from inexpensive mobile phones to high-end PDAs and smartphones. Also, they are available in a variety of sizes and provide different functionality and performance.

Some mobile devices are used extensively for entertainment such as gaming; others can be used for watching movies and listening to music; and some are used for communication in the form of email, for example. Whatever the usage of the



J. Téllez, S. Zeadally, *Mobile Payment Systems*, Computer Communications
and Networks, DOI 10.1007/978-3-319-23033-7_2

device, common characteristics like portability, small display, limited processing power, and limited battery remain the same. Therefore, based on these features, the devices can be classified into the following seven classes (see Fig. 2.1) [110, 122]:

- *Notebooks:* A notebook is a small and portable computer that normally sits somewhere between laptops and PDAs in terms of both size and functionality. Usually notebooks do not come with a full keyboard but they often include extra characteristics like a touchscreen. Also, most of notebooks execute standard personal computer operating systems whereas other devices execute more lightweight operating systems.
- *Tablets:* A tablet computer, or simply a tablet, is a type of internet-enabled computer that works in a similar way to a smartphones. Tablets have an operating system similar to laptops and are equipped with cameras, accelerometers, and touch screens that need to be operated with a stylus, digital pen, or fingertip instead of a keyboard or mouse. They are slim and lightweight, which makes them more portable than a standard laptop.
- *Mobile Media Players:* A mobile media player is a mobile device specifically designed for accessing multimedia content, like music and videos, on the go. Some of them feature wireless connectivity hardware for accessing content through a network, and high-end devices often include portable video players and recorders. Most mobile media players are equipped with their own operating systems (except high-end devices that often run common operating systems), which do not allow the installation of extra software.
- *Mobile Gaming Devices:* A mobile gaming device, or mobile entertainment device, is a lightweight mobile device with a built in screen, game controls, and speakers, principally designed for playing computer games. Moreover, most of these devices can play multimedia content, and newer devices feature wireless connectivity to support multiplayer games. Also, they have their own personalized operating system that only allows the installation of games.
- *Mobile Phones:* A mobile phone (also known as a cellular phone, cell phone, or hand phone) is a device that basically can make and receive telephone calls as well as send and receive text messages while moving around a wide geographic area. Mobile phones run a very simple operating system and some of them offer features like synchronization of contacts or the calendar with a computer (desktop or laptop).
- *Personal Digital Assistants (PDA):* A PDA, also known as a palmtop computer, is a mobile device that is equipped with a fast processor, large touch screen, built-in audio device, common operating systems allowing the installation of additional software, wireless connectivity, and so on. They were used to manage daily tasks like meetings, appointments, reminders, and contacts, as well as to perform complex tasks.
- *Smartphones:* A smartphone is a mobile phone that has the ability to run local applications without requiring network coverage. Also, it provides a standardized interface and platform for application developers by running its own complete operating system.

**Fig. 2.1** Examples of mobile devices

## *2.1.2 Mobile Operating Systems (Mobile OSs)*

A mobile operating system, also referred to as a mobile OS, is an operating system that operates a smart- phone, tablet, PDA, or other mobile device and provides a software platform to run application programs. There are several operating systems (OSs) available in the market based on the requirements of the mobile device, each one with its advantages and disadvantages. These OSs are usually different from the OSs developed for personal computers due to the specific hardware and particular requirements that create new constraints related to energy and space.

Traditional mobile operating systems are lightweight operating systems that run on minimal resources such as low memory, low processing power, for example. Modern OSs are fast, perform multitasking (such as gaming and making phone calls at the same time), and combine the features of a personal computer OS with other features, including a touchscreen, GPS mobile navigation, camera, video camera, speech recognition, voice recorder, music player, short range link (such as Bluetooth, Wi-Fi, NFC), and so on. However, the major constraints faced by mobile operating systems include a small display and the lack of a full keyboard as compared to desktop OSs, which have high performance internal hardware and a more friendly user interface [110].

Many OSs have been developed for mobile device. However, only the most common OSs for existing consumer-type mobile devices are presented here.

- *Symbian OS:* Symbian OS, a descendant of Psion's EPOC [144] ] that runs on ARM processors, was developed by Symbian Ltd. with the associated libraries, user interface, frameworks, and reference implementations of common tools.

It was one of the most popular OSs used by major mobile device manufacturers including Nokia, LG, BenQ, Samsung, Motorola, and Sony Ericsson. Symbian OS was the most popular smartphone OS based on a worldwide average until the end of 2010, when Android overtook it.

Symbian OS requirements regarding CPU power and storage space are low, and it is very optimized for saving battery power. Its design patterns include using the kernel to the minimum, sharing resources between users, having GUI-based applications, capturing user interaction as events and making them available in the form of event queue, multi-tasking and multi-threading, having persistent data storage, and providing user services and access to kernel services in the form of class libraries [122].

- *Palm OS:* Palm OS is a mobile operating system initially developed by Palm, Inc. for its PDAs, but later licensed to other device manufacturers, such as Sony and Handspring. It was designed for ease of use with a touchscreen-based graphical user interface. In the Rapid Storage Technology (RST) versions, Palm OS was designed to run on very limited hardware without providing features like multi-processing or multithreading. However, later versions were extended to support smartphones, introducing additional features like improved multi-processing support and built-in file system encryption.

  Some of the features Palm OS offer in the latest version are: a single tasking environment to allow full screen applications, a handwriting recognition scheme called Graffiti 2, synchronization with desktop computers using HotSync, a simple security model, TCP/IP access, and a serial port/USB, Infrared, Bluetooth, and Wi-Fi connections.
- *Windows Phone (WP):* WP is a proprietary mobile operating system developed by Microsoft for smartphones and mobile devices. It is the successor to Microsoft's initial mobile OS platform system (called Windows Mobile), which introduced a new design language called Modern (formerly known as Metro UI) that was inspired by the user interface in the Zune HD. The latest release is WP8, which was launched in October 2012. Metro has a home screen (called the start screen) that is made up of "Live Tiles", which inspired the Windows 8 live tiles. Tiles are links to application, features, functions, and individual items such as contacts, web pages, or media items [188, 190]. WP 8 supports real multitasking technology that allows developers to create apps that can run in the background and resume instantly. Also, to allow applications to be easily ported between WP 8 and Windows CE-based, WP 8 replaced its previously Windows CE architecture with one based on the Windows NT kernel with many components shared with Windows 8.
- *BlackBerry OS:* BlackBerry OS is a proprietary multitasking environment developed by BlackBerry Ltd. (formerly Research In Motion Limited or RIM) for its BlackBerry line of smartphone handheld devices. The operating system enables multitasking and supports specialized input devices that BlackBerry Ltd. has adopted for use in its handhelds, particularly the trackwheel, track-ball, and, most recently, the trackpad and touchscreen. The BlackBerry OS provides support for Java MIDP 1.0, WAP 1.2, a subset of MIDPS 2.0 (after version 5.0), and allows

complete wireless activation and synchronization with exchange email, calendar, tasks, notes, and contacts. It also adds support for Novell GroupWise and Lotus Notes. BlackBerry OS was discontinued after the release of BlackBerry 10, but BlackBerry will continue support for the BlackBerry OS.

- *Linux:* Linux, as a mobile device OS, has a long history in the mobile world since many mobile devices have been developed with Linux-based operating systems. Linux can be used in different ways but the kernel and the system libraries remain the same across various devices. Also, since Linux does not offer a generic user interface system, every manufacturer selects one of the existing systems or develops one.

  Despite the advantage offered by Linux with respect to other mobile devices, OSs, it requires a large memory for runtime and storage, and does not have sophisticated power-saving mechanisms [122].
- *iOS:* iOS (previously iPhone OS) is a mobile OS developed and distributed by Apple Inc. for a number of Apple devices including the iPhone, iPad, iPad 2, and iPod Touch. It is Apple's mobile version of the OS X used on Apple computers which uses Objective C (based on C language) for its touch Software Development Kit (SDK) and takes about 1.5 GB of the device's total memory storage. The user interface of iOS is based on the concept of direct manipulation using multi-touch gestures. Interaction with the OS includes gestures such as swipe, tap, pinch, and reverse pinch, all of which have specific definitions within the context of the iOS and its multi-touch interface.
- *Android:* Android, one of the famous current OSs, is an open source mobile OS developed by Google, Inc. Android is an OS based on the Linux kernel and designed primarily for touchscreen mobile devices such as smartphones and tablet computers. Application programming is exclusively done in Java, and the OS allows the installation of third-party applications. The Linux kernel is responsible for executing core system services (such as memory access, process management, access to physical devices through drivers, network management, and security) whilst the Linux kernel is the Dalvik virtual machine, which is a register-based execution engine used to run Android applications [33].

Figure 2.2 shows a graphical representation of the market share of each of the mobile phone OSs available [50].

## 2.2 Mobile Device Security

In recent years, mobile devices have enabled access to a large variety of ubiquitous and mobile services due to the significant increase in the different forms of connectivity mobile devices provide, such as GSM, GPRS, Bluetooth, and Wi-Fi. The power of today's mobile devices is sufficient to consider them as personal computers because they can be used to do most of the things that people do on a desktop, including processor-intensive tasks such as watching movies. Therefore,

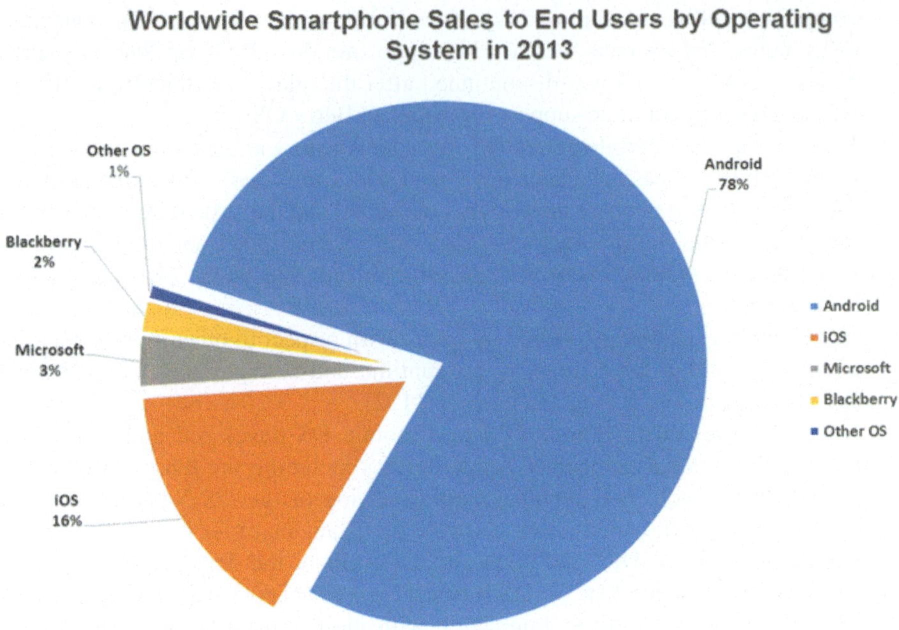


**Fig. 2.2** Market share of mobile phone OSs in 2013 [50]

it is clear that mobile devices have rapidly become a critical part of everyday life. However, since these devices hold valuable, sensitive content, and possibly classified information, they face the same or higher levels of attacks and threat of privacy violation that have affected the desktop computing environment. Also, the fact that mobile devices can be carried wherever people go increases the risk of losing them and their content. If a mobile device is lost or stolen, confidentiality is broken. Even if the device is found after a few days, the confidentiality of the device has been lost. Once a mobile device is compromised, an attacker can perform one of the following malicious operations: (a) selling personal data that has been uncovered, (b) gaining access to a device owner's accounts by using stored credentials, (c) using the device as a gateway into enterprise data and resources by leveraging a trust relationship between the device and the IT infrastructure, and (d) using the device to send unauthorized premium-rate SMS messages or unauthorized email [102, 142, 184].

According to a survey carried out by ESET Latin America on the ***uses made of mobile devices***, we can understand that, although storage of private information and passwords is not the primary use of mobile devices at the moment, it does constitute a notable percentage of common tasks [38]. Figure 2.3 shows the relevant statistics from the aforementioned survey.

Since mobile devices are becoming critical to our lives, it is necessary to know how to make them robust and secure.

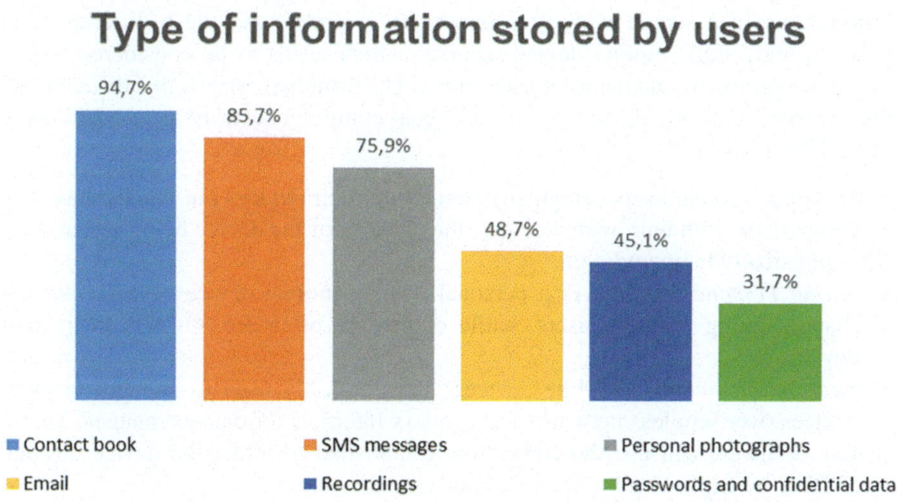


**Fig. 2.3** Type of information stored by users in mobile devices [38]

### *2.2.1 Mobile Device Security Versus Personal Computer Security*

In spite of the similarities of smartphones and personal computers (PCs), three important differences between mobile device security and personal computer security exist.

- Firstly, from illicit activities, malware authors can generate costs for the user and revenue for themselves easily on smartphones; unlike desktop environments where the emergence of flat rates and DSL have eliminated the possibility for a malware to dial premium rate numbers due to the absence of a telephone connection. For example, premium-rate calls or SMS attacks could feasibly go unnoticed until the user's next phone bill [39].
- Secondly, the presence of the mobile network operator and its influence on the device. In contrast to PCs, where the network provider almost always has no influence on the user's computer, the mobile network operator has a trusted device inside the mobile phone (the SIM card). It is possible to create trusted applications on the mobile phone with enhanced security in combination with the SAT [12].
- Thirdly, the mobile network operator will invoice any cost-generating smartphone activity even when malware may have generated it. Therefore, users can expect that the mobile network operator will charge them even if malware is responsible for the costs.

Security issues on mobile devices differ from security issues on PCs and servers. Since in an MPS customers have a mobile device to perform e-commerce transactions in a secure way, it is important to understand these differences in order to understand mobile device security, which needs to be considered before designing protocols and deploying solutions. The principal aspects that differentiate the security of mobile devices from traditional computer security are the following [122, 129, 142]:

- *Mobility:* Mobile users can go anywhere with their devices but this increases, in comparison with unmoving devices, the chances of the device being stolen, lost, or physically tampered with.
- *Strong Personalization:* As a personal device, mobile devices usually are not shared among multiple users, while computers often are. Owners keep their devices close.
- *Strong Connectivity:* Mobile devices are commonly used to connect to other devices over wireless networks (or wireless Internet) for data exchanges. Therefore, malware can use the connection established to infect the device through SMS or by exploiting the Internet connection.
- *Technology Convergence:* Today numerous functional features are integrated in the mobile device (such as gaming, video and data sharing, and internet browsing) which can be used by attackers to explode various routes to execute their attacks
- *Limited Resources and Reduced Capabilities*: Compared with stationary devices, mobile devices have four major limitations: (a) limited battery life, (b) limited computing power, (c) very small display screen size, and d) very small keys for inputing. These limitations create challenges in building mobile security technology.

For Mulliner [122], the aforementioned aspects are some of the reasons to claim that complexity of the security of mobile devices is greater than the security of traditional computers. For example, *mobility* raises the risk of data theft because it is much easier to steal a mobile device, than it is to attack a computer. Moreover, *strong personalization* along with *strong connectivity* raises the menace of privacy violations. On the other hand, *technology convergence* leads to additional security risks due to the fact that each extra characteristic creates one or more novel subjects to be attacked. *Reduced hardware* capacities could help some type of Denial-of-Service (DoS) attacks. Additionally, the absence of some characteristics like a full keyboard, makes it hard to implement efficient authentication mechanisms (e.g., username and password). All these aspects establish extra implications such as the increase of complexity when a security audit on mobile devices is performed.

### 2.2.2 Threat Model

The knowledge of the existing kinds of threats to the system allows for the improvement of its security. A model which identifies the threats against systems, the resources to be protected, the features of the attackers, and the potential

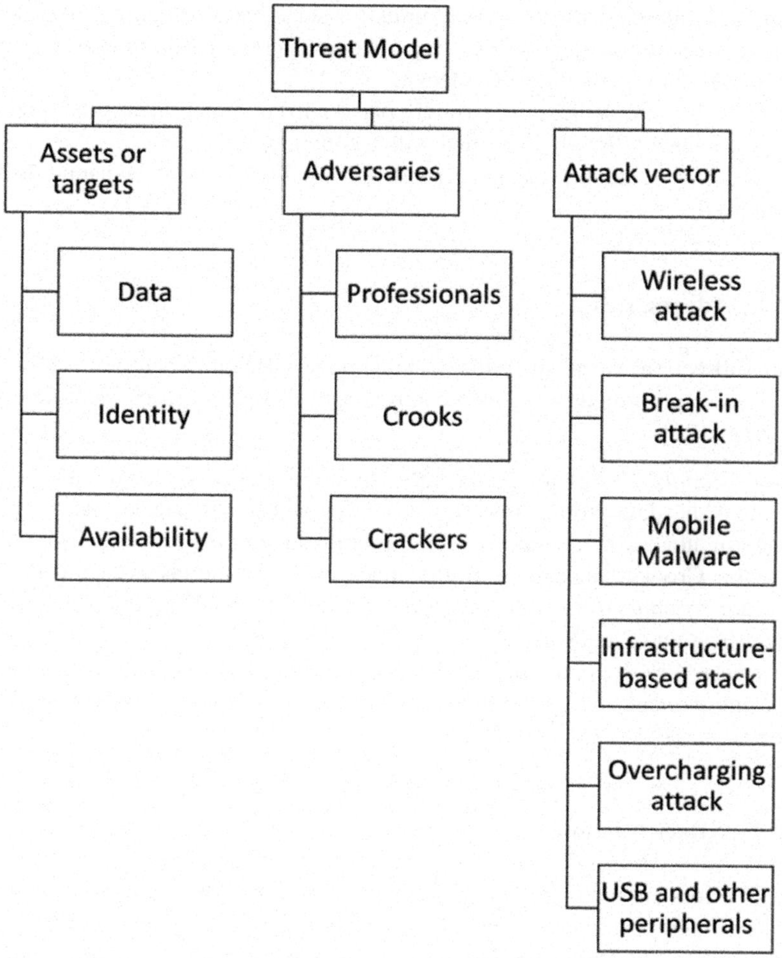


**Fig. 2.4** Threat model for mobile platforms

attack vectors is known as *Threat Model* (see Fig. 2.4). However, it is important to understand that issues of traditional computer security (such as *confidentiality*, *integrity*, and *availability* [122]) are the same as those faced by mobile device security deals.

#### 2.2.2.1 Targets for Mobile Device

In this subsection, we identify the following targets of mobile devices that must be protected to ensure the system's security:

- *Data:* Mobile devices are storage units for sensitive information, like authentication credentials, activity logs (e.g., phone usage), and commercial or private information (e.g., pictures or videos).
- *Identity:* Mobile devices are strongly personalized, which means that the device or its content is directly identified with a concrete user.
- *Availability:* Availability is not considered a real "asset" since it cannot be stolen or abused.

#### 2.2.2.2 Adversaries

In this subsection, we identify the possible adversaries that can perform an attack on the assets presented in the above subsection. The adversaries identified are the following [122]:

- *Professionals:* Professional attackers use the identity associated with the device to steal the data stored on a device or for additional attacks. Also, they can perform attacks on the three assets of mobile devices mentioned above.
- *Crooks:* Crooks concentrate on the attack of all the targets as they can in order to gain revenue from the data store in a device which is associated with the identification of a person.
- *Crackers:* Crackers are interested in the creation of viruses and worms or in just causing damage. Also, they may have interest in obtaining specific data from a device.

#### 2.2.2.3 Attack Vectors

Attack vectors are paths through a security concept by which a hacker (or cracker) can elevate permission levels and gain access to the system or data, which are stored on the system. The following attack vectors can be identified [33, 122]:

- *Wireless Attack:* Personal and sensitive data stored in mobile devices are targets of many types of wireless attacks. Eavesdropping is the most usual attack against wireless transmissions to extract confidential information, such as usernames and passwords.
- *Break-in Attack:* In break-in attacks, attackers manage to gain control of the target in a partial or full manner. It exists in two forms: (a) *code injection*, achieved through exploitation of programming errors, which lead to buffer overflows or format string vulnerabilities, and (b) *abuse of logic errors*, which attack logic errors that are specific to the application or device. The confidentiality, integrity, and availability of a device are impacted by the break-in attacks.
- *Mobile Malware:* This is any kind of hostile, intrusive, or annoying software or program code designed to use a device without the owner's consent. In Sect. 2.2.3, we discuss in detail this kind of attack.

- *Infrastructure-based Attack:* The infrastructure based attacks mainly target GSM networks and the application server of the mobile device. Any threat to these areas is a major issue as these provide the basis for primary mobile device functionalities such as phone functionality. Because these attacks can be considered part of one or more of the attacks categories, they must be treated in a explicit manner due to their interaction with the mobile infrastructure.
- *Overcharging Attack:* These attacks are very specific to wireless mobile devices and involve a paid service of some kind (e.g., a mobile phone service agreement) where the objetive of the attacker is to charge additional fees to the victim's account and then to try to transfer these extra fees (money/credits) from the victim to the attacker.
- *USB and Other Peripherals:* The USB connection is commonly used to synchronize the mobile device with a personal computer. Therefore, an attacker can try to compromise the software used to synchronize the mobile device to access private information and install malicious applications on the device. Also, since some mobile platform allows a device to be connected as a USB storage device, attackers could perform traditional USB malware attacks.

### *2.2.3 Mobile Malware*

With the increasing improvement of mobile device capabilities, their functionalities are becoming more similar to computers, and the mobile platforms increasingly resemble traditional OSs. This will cause a continual increase in the popularity of mobile devices, but the security threats typical to PCs will migrate to mobile devices, causing an increase in the trend of attacking mobile devices by infecting applications with malicious content [33, 140].

*Malware* is software with a malicious purpose. It can be defined as any type of hostile, intrusive, or obtrusive software or program code (e.g., trojan, rootkit, backdoor, etc.) designed to control a device without the consent of the owner. Generally it is delivered through a spam email in the form of a malicious attachment or a link to an infected website [142]. Since mobile malware comes in many forms, the popular mobile malwares are described briefly here:

- *Virus:* A virus is a portion of program code that, when executed, replicates by inserting or attaching copies of itself (possibly modified) into other computer programs, data files, or the boot sector of the hard drive.
- *Worm:* A worm is malicious program that self-replicates to propagate autonomously to systems that have not been infected across the usage of various transport mechanisms through an existent network in the absence of the intervention of the user. Ikee.B represents a worm-type malware for mobile devices utilized on jailbroken iPhones to rob financially sensitive data [42].

- *Trojan:* A trojan is a software that masquerades as something it is not, appearing to provide some functionalities while containing a malicious program. Trojans are often used to commit phishing activities or to install other malicious applications like worms or botnets.
- *Rootkit:* A rootkit is a malicious application that infects an operating system to gain rights to run itself in a privileged mode. Usually, rookits mask their presence from the user by modifying standard OS functionalities (to operate stealthily and retain longer control over the infected devices) or by installing trojans or disabling firewalls and anti-virus software.

  Recent research efforts indicate the potential of rootkits's attacks strategy and considered them to be an emerging threat to mobile security [16].
- *Bonet:* A botnet is a network of compromised devices (infected by a virus) that an attacker can control and coordinate remotely. Since the attack strategy is to employ the computing power of compromised devices to perpetrate different activities varying from sending spam email to committing DoS attacks. Waledac is a specific botnet for mobile devices that employs SMS and MMS messages to swap data among nodes and keep the botnet alive in all the nodes even if they have no Internet connection [33].

Mobile malware is a real threat and can severely impact end users, either financially or emotionally. They can use different vectors to propagate, like an SMS that contains a link that force users to download a malicious code from a website, infected attachments included in an MMS, or infected programs received through NFC or Bluetooth. However, in the context of smartphones, malware are designed to steal sensitive data saved in the phone or the user's credit card information [142]. Hence, it is important to install anti-virus and security applications to protect mobile devices against mobile malware attacks. However, those applications can only detect previously identified mobile malware and will not detect newly designed malware. Therefore, users should take additional steps in order to protect their mobile devices against malware infections [140]. Töyssy et al. [179] propose some prevention solutions and countermeasures that will make malware prevention feasible by considering:

- *Users' education*, which helps users to utilize the mobile device in a secure way.
- *Software developers*, which can develop security protection designed for smartphones such as antivirus software.
- *Network operators*, which can use mechanisms to avoid intrusions by enhancing their network infrastructure.
- *Phone manufacturers*, which should update the devices automatically (without cost to the user) to keep phones up to date, thereby making it more difficult for attackers to exploit security holes.
- *Epidemiological models*, which can predict whether a virus that has been detected will turn into an epidemic.

For a comprehensive discussion of similar solutions to protect mobile devices against malware infections, see [186] and [140].

## 2.2.4 *Security Solutions for Mobile Devices*

This section presents, in two categories, the major current mechanisms developed to avoid various kinds of threats for smartphones: Intrusion Detection Systems (IDSs) for smartphones and trusted mobile-based solutions.

### 2.2.4.1 Intrusion Detection Systems

An IDS is a device or application that dynamically monitors the system's and the user's actions in order to detect intrusions. Since an information system can suffer from various kinds of security vulnerabilities, it is both technically difficult and economically costly to build and maintain a system that is not susceptible to attacks. Therefore, by analyzing the system and user operations in search of undesirable and suspicious activity, IDSs can effectively monitor and protect against threats. IDSs can be based on two complementary types of approach [170]:

- *Prevention-based* approaches, which use cryptographic algorithms, digital signatures, hash functions, and important properties (such as confidentiality, authentication, or integrity), can effectively reduce attacks by ensuring that users conform to predefined security policies. In this scenario, IDSs have to be running online and in real-time.
- *Detection-based* IDSs serve as a first line of defense by effectively identifying malicious activities.

On the other hand, two main types of detection can be identified:

- *Anomaly-based:* Also referred to as anomaly detection, or behavior-based, anomaly-based detection observes a deviation from normal or expected behaviors of systems or users to detect an intrusion. In this approach, alterations of recognized attacks or new attacks can be detected by IDS but the disadvantage is that the quantity of false alarms is generally pretty high.
- *Signature-based:* Also referred to as signature detection, misuse-based, knowledge based, and detection by appearance, signature-based detection is based on a database of well-known attack signatures and system vulnerabilities and tries to identify activities matching a signature that is stored in the database. The advantage of this approach is that the number of false alarms is generally very low but the drawback is its ability to identify only known attacks.

The aforementioned types of detection can be combined to create hybrid approaches. Their effectiveness is measured using the following metrics: true positive rate, accuracy, and response time.

#### 2.2.4.2 Trusted Mobile

Due to the trend in recent years of mobile devices replacing desktop PCs as the main computing platform for many users, numerous organizations are beginning to allow employees to access enterprise resources from their personal mobile devices. However, as stated in Sect. 2.2.1, most mobile devices lack the security measures available in more traditional computing platforms, and enterprises are concerned about the risks associated with integrating mobile devices into their networks. Mobile computing is exposed to new vulnerabilities and attacks that slow down the evolution and widespread adoption of new usage models on mobile platforms.

The National Institute of Standards and Technology (NIST) suggests a set of desired capacities for security-enabled mobile devices [176]: (a) *device integrity*, which is achieved when its software, firmware, and hardware configurations are in an expected state; (b) *device isolation*, which provides divided domains for computations and data owned by various entities, or stakeholders, on the same device; and (c) *protected storage*, which allows mobile devices to preserve the confidentiality and integrity of data on the device when in use or resting.

A mobile device should be capable of providing evidence of device integrity to a third party by some form of device attestation. An attestation is described as a request made by a device (called the attester) to a third party (called an appraiser) regarding its properties and providing proof to support that claim. The appraiser makes a decision based on the provided evidence. A common device attestation is depicted in Fig. 2.5, where two devices can be identified: (a) the attester, which is a device that seeks access to some protected resource, and (b) the appraiser, which is a remote server that acts as a gatekeeper for the protected resource. The process starts when the device solicits access to the resource. Then, as the appraiser requires a device attestation, the mobile device will provide some measurement data that represent the state of the device as evidence of its integrity. Upon receipt of this evidence, the appraiser evaluates it to determine whether the device is in an expected state. If the appraiser is satisfied, the attestation passes, and the appraiser grants access to the resource [115].

To address the new security challenges and new security requirements of mobile devices, the Trusted Computing Group (TCG) formed a working group dedicated to the mobile phone area that has published a set of specifications to measure, store, and report hardware and software integrity through a hardware *root-of-trust*, which is the Trusted Platform Module (TPM) and Core-Root-of-Trust-Measurement (CRTM). A TPM can be defined as a discrete hardware component of a platform that provides secure generation and management of cryptographic identities, isolated processing resources, and protected storage for sensitive data. The CRTM on a TPM-enabled platform measures the bootloader of the system before it is executed and then stores the measured value into one of the Platform Configuration Registers (PCRs) inside the TPM. Then, the bootloader loads the OS image, measures it, and stores via a PCR extension before executing it. The TPM assigns a set of PCR values with an Attestation Identity Key (AIK) after an attestation challenge from a third party and then sends back the result. The challenger then can make decisions on the

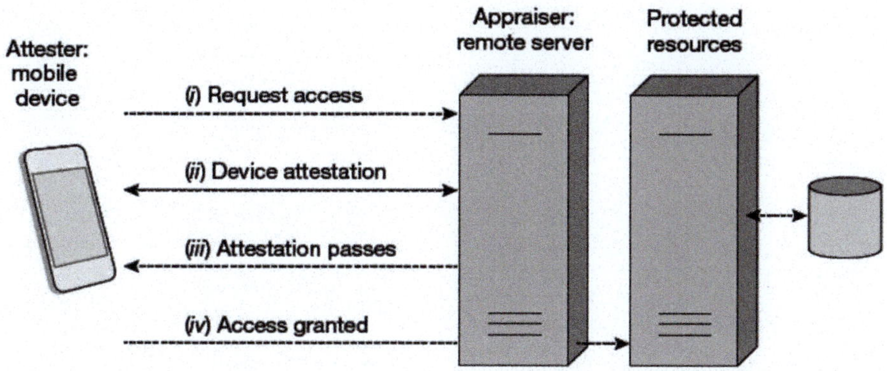


**Fig. 2.5** Device attestation [115]

trust status of the platform by verifying the integrity of these values and comparing them with the corresponding known good values [115, 198].

Mobile Trusted Module (MTM) is a specification released by the TCG Mobile Phone Working Group for mobile phone platforms. MTM could increase the security of smartphones by providing basic cryptographic capabilities (such as random number generation, hashing, protected storage of sensitive data, asymmetric encryption, and generation of signatures) that can be exploited to implement hardware- based security services, such as device authentication, integrity measurement, secure boot, and remote attestation. Also, the MTM provides a root-of-trust for smartphones in the same way that the TPM does for personal computers [142].

# Chapter 3
# Architectures and Models for Mobile Payment Systems

In the last decade, hundreds of mobile payment models have been proposed around the world. These models may be classified according to the type of payment made, based on the technology adopted to implement the solution, the cryptographic technique used to secure the communication, and so on. However, in this chapter we present a classification focused on the features and technology most apparent to the user. The purpose of our classification is to help readers easily find the state of the art of mobile payment models found in the literature based on their core features. Figure 3.1 shows our classification of mobile payment models.

## 3.1 Classifications of Mobile Payment Models: State of the Art

### *3.1.1 Micro-payments*

Micro-payments are low-value financial transactions that vary from pennies to a few dollars that are made very quickly, like paying for a ringtone download. Given these constraints, micro-payment techniques must be both inexpensive and fast [9].

Despite of the transaction amount in micro-payments being small, there exist a large number of users and transactions. Therefore, an unsafe transaction of even just a small percentage can mean a substantial loss of income. Thus, security is an important issue of micro-payments [67].

Many micro-payment systems that work very well for the Internet can be found in the literature. All of these can be classfied into script-based (such as Millicent [52, 112], Subsrip [45]), hashchain-based (such as PayWord [149]), and macro-payment-based (such as μ-iKP [13, 14, 63]) categories. However, they are not adequate for a mobile data network owing to the restricted bandwidth of the



J. Téllez, S. Zeadally, *Mobile Payment Systems*, Computer Communications and Networks, DOI 10.1007/978-3-319-23033-7_3

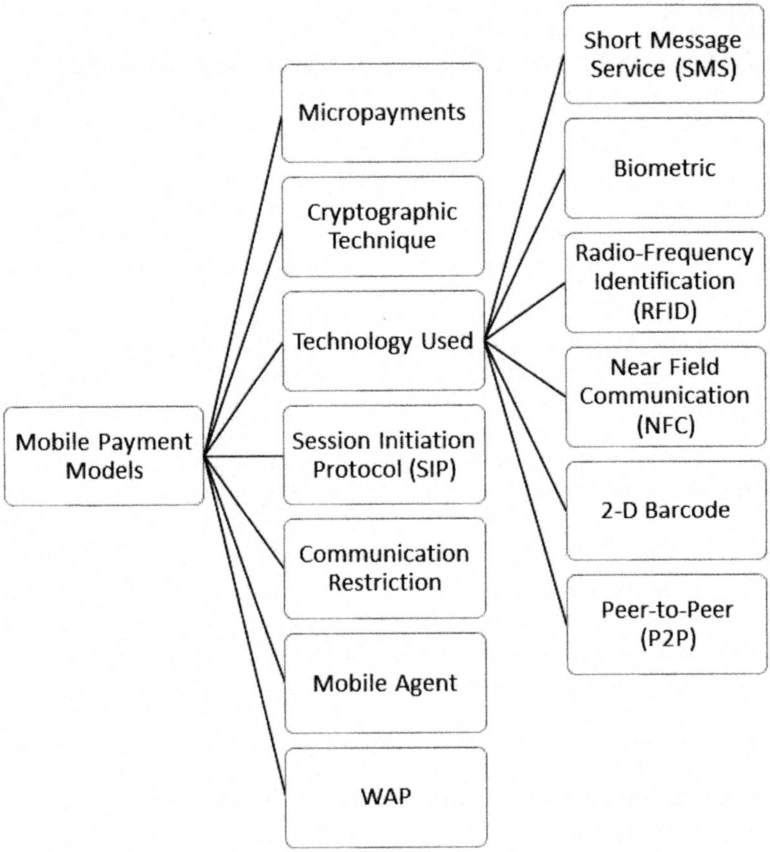


**Fig. 3.1** Classification of mobile payment models

mobile network, and the constraints of mobile phones/smartcards regarding the computational capabilities and memory resources.

A new secure mechanism for mobile micro-payments, Anonymous Micro-payments Authentication (AMA), was proposed by [67]. This allows customers and merchants to authenticate each other indirectly (without revealing the customer's real identity to the merchant) using the proposed model shown in Fig. 3.2. In the proposal, a customer is able to receive fast micro-payments both from his/her local domain and from a remote domain keeping the load on their mobile phone/smartcard. Additionally, as the communication overhead is not increased, computational overheads on the mobile phone/smartcard is reduced.

Another protocol proposed in 2004 to support m-commerce transactions was MobiCoin (based on PayCash [137], an e-commerce protocol). MobiCoin was proposed by [2] and designed to work offine, without involving a third party broker for every transaction. This change makes it all the more efficient because the communication overhead previously needed to contact the third-party is eliminated. On the other hand, the advantages of PayCash are included in the proposal.

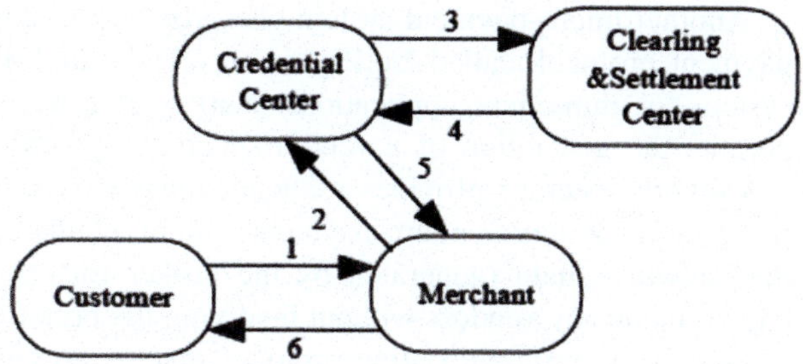


**Fig. 3.2** AMA model [67]

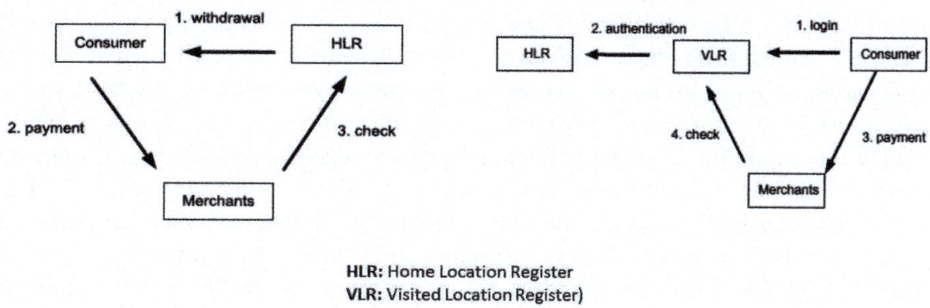


**Fig. 3.3** The novel online mobile payment scheme payment processes: payment in home domain (*left*) and payment in visited domain (*right*) [68]

The proposed protocol employs SIM cards for data protection and active certificates for distributed trust. MobiCoin is secure, durable, fair, atomic, and accountable. It may be used as a digital cash protocol with a mobile digital wallet without the trade-off of anonymity. In addition, it is an online protocol, thus increasing the efficiency and availability of m-commerce transactions.

MobiCoin eliminates the need for a centralized agency during payment making the system both fast and inexpensive. Moreover, MobiCoin is secure in terms of preventing customer fraud, double-spending, and non-repudiation, and it allows for partial and full anonymity. Also, the protocol supports micro-payments with the same proposed scheme.

A novel online mobile payment scheme that supports roaming service was proposed by [68]. The proposed scheme can be performed in both the home domain and in the visited domain (see Fig. 3.3). Moreover, the scheme provides consumer anonymity, authentication, non-repudiation, data integrity properties (using MAC technique), and double-spending prevention.

The proposed scheme can deal with different types of merchants, such as shops, vending machines, and WAP sites. Moreover, the scheme only uses a one-way hash function and symmetric encryption, so it has high performance and can be applied easily to a mobile device that has low computing ability and less memory storage. Also, the scheme is practical and efficient in mobile e-commerce.

Another micro-payment system was proposed in 2007 by [31]. This new micro-payment protocol, called NetPay, is an off-line and decentralized micro-payment system for thin-client applications. NetPay is a payword-based protocol (which permits the generation of a vendor-specific payword chain that can be spend at that vendor-specific), provides a safe, inexpensive, widely available, and debit-based protocol for an online micro-payment system. Unlike payword protocol, in NetPay the payword chain (generated by the broker for each customer who can spend paywords on any vendors without involving the broker) is not vendor specific.

NetPay is a simple online protocol that secures the anonymity of customers from vendors whilst helping them to avoid double-spending. Moreover, NetPay is secure against attack from both internal and external adversaries. The proposed protocol varies from those former protocols based on payword in that it uses touchstones signed by the broker and an e-coin index signed by vendors, which are transmitted from vendor to vendor. A vendor utilizes the signed touchstone to verify the electronic currency paywords, and the signed index is used to impede double-spending of customers and to solve conflicts among vendors [31]. The broker (responsible for maintaining the accounts opened by customers and vendors and the funds deposited by them) is asummed to be honest and trusted by customers and vendors, who could be dishonest. Figure 3.4 shows the NetPay payment model.

In order to compare the features of the proposed NetPay protocol with other micro-payment protocols, the authors depict a design and prototype implementation of NetPay for deployment with thin-client (such as PDAs, mobile phones, etc.) vendor-user interfaces for customers. Two example applications that include NetPay micro-payment support are included, a purpose-built e-newspaper and a pre-existing component-based e-journal web site. Also, three kinds of evaluations on the proposed NetPay prototypes are presented in the work to prove their usability, performance effect on a vendor's e-commerce system, and general satisfaction based on the following requisites: an easy-to-use and robust micro-payment system;

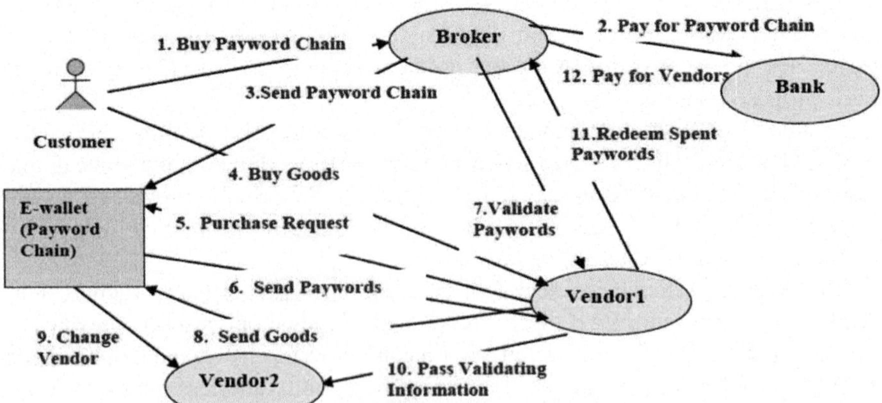


**Fig. 3.4** NetPay basic interaction between the parties [31]

off-line processing support; secure e-coins and no double-spending; anonymity for customers; transferable e-coins among vendors; and a scalable micro-payment architecture that can support a large number of customers and low performance impact. Despite the advantages of the approach, the transmission of the touchstone and an e-coin index of a customer from one vendor to another could fail if one vendor system goes down.

In 2008, [126] proposed a digital coin system featuring both observer[1] and fairness (that implement an anonymity revocation mechanism to prevent criminal activities or to trace the criminals), showing that both concepts do not interfere with each other and can be implemented simultaneously without loss of security. Unlike other fair off-line coin systems (those in which the vendor has to check the validity of coins without contacting the bank), fairness is implemented with the help of the observer, thereby reducing the computational effort during the withdrawal.

The proposed digital coin system provides a physical defense as primary protection against double-spending by using a dedicated tamper-resistant device, supplemented by a well-known double-spending detection mechanism. The approach is to modify a coin system with observers so that a Trusted Third Party (TTP) can revoke the anonymity of customers if necessary. Since crucial parts of the coins are stored by the observer and cannot be read by their owner, it is not obvious how to design a coin system that simultaneously provides both physical defense (by an observer) against double-spending and the fairness of coins.

An efficient and practical protocol to carry out micro-payments (based on the use of anonymous mobile cash) that provides anonymity and inability to link to customers was proposed in [113]. The authors of the proposal have developed an alternative withdrawal process, which is secure against any internal or external attacks. The cash is withdrawn only if the customer is a legitimate user of the system and has sufficient money in his/her bank account. The customer cannot use m-cash (mobile cash) without the stamp the bank issuer provides.

The proposed mobile payment was designed for real point of sale scenarios, providing anonymity to customers during the payment phase. Moreover, this protocol prevents double-spending and forgery attacks. Also, the protocol requires a low computational cost.

A scheme to enable micro-payments efficiently in SEMOPS[2] architecture was proposed by [53]. The proposed scheme has been fitted in SEMOPS architecture with no significant modifications to the original architecture. Unlike many other micro-payment mechanisms, the suggested approach makes use of intermediaries to aggregate micro-payments, thus achieving efficiency (aggregating micro-payments forms a macro-payment, and the macro-payment amount is paid, thus reducing the overall cost per payment).

[1]A dedicated tamper-resistant device that is used to physically prevent illegitimate copying of coins.

[2]SEMOPS is a secure mobile payment service proposed by [185] that addresses the major roadblocks in developing a mobile payment infrastructure: standardized protocols, interoperability, and security. However, it will work inefficiently for micro-payments due to of the need for significant computation and communication for every payment.

The proposed approach does not put the burden of any computationally complex operation on the mobile phone, which is very important in a mobile commerce environment. The micro-payment protocol developed considerably reduces the amount of computation and number of messages passed per micro-payment. This reduces the total cost of each micro-payment.

In 2009, a new mobile micro-payment protocol based on the chaos hash function and symmetric encryption algorithm was proposed by [125]. This protocol generates a pay certificate and encrypts order messages transmitted in the network with a symmetric encryption algorithm, which has greater security and efficiency and is suitable for those micro-payment transactions that have very small trade amounts and only require partial fairness.

The new protocol addresses the deficiency of having no necessary anonymous character and partial fairness in the PayWord protocol and ensures information confidentiality. It adopts the security chaos hash function to replace the standard hash function of MD5 and SHA, which suffer from collision danger.

In the same year, [10] presented a scheme that support m-commerce transactions with complete anonymity for mobile users. Based on the e-commerce protocol DigiCash payment system [88, 134], MobiCash is devised to work online for mobile devices that support security, efficiency, and full anonymity for mobile users. The proposed scheme is based on 160-bit ECC, which ensures the same security strength to 1024-bit RSA.

MobiCash is an online MPS based on an electronic wallet (containing digital coins) that provides to a mobile user a high level of privacy and anonymity. Additionally, the proposed protocol is capable of preventing double-spending and overspending of coins.

Before any transaction takes place, it is necessary to establish a secure channel between a TTP and banks (both the customer's bank and the merchant's bank). Also, the TTP should generate a digital certificate for the banks in order to verify the banks' signatures. The steps of the proposed protocol are as follows (as is presented in Fig. 3.5) [10]:

I. Before ordering items, the customer visits the merchant's web site to choose them.
II. The merchant creates a *pay_req* message (which holds information regarding the order, amount, the currency to be used, and the current time) and then sends it to the customer's electronic wallet.
III. Upon receiving the *pay_req* message, the customer signs it and generates a *signed_pay_req* which they send to the merchant.
IV. When the merchant receives the *signed_pay_req* message, they create the *pay_req_response* message and send it to the customer.
V. Upon receiving the *pay_req_response* message, the coins are generated by the customer with the use of the electronic wallet installed on the mobile device. Then, a blind signature algorithm is applied to coins by the customer before sending the *withdraw_req* message to the customer's bank.

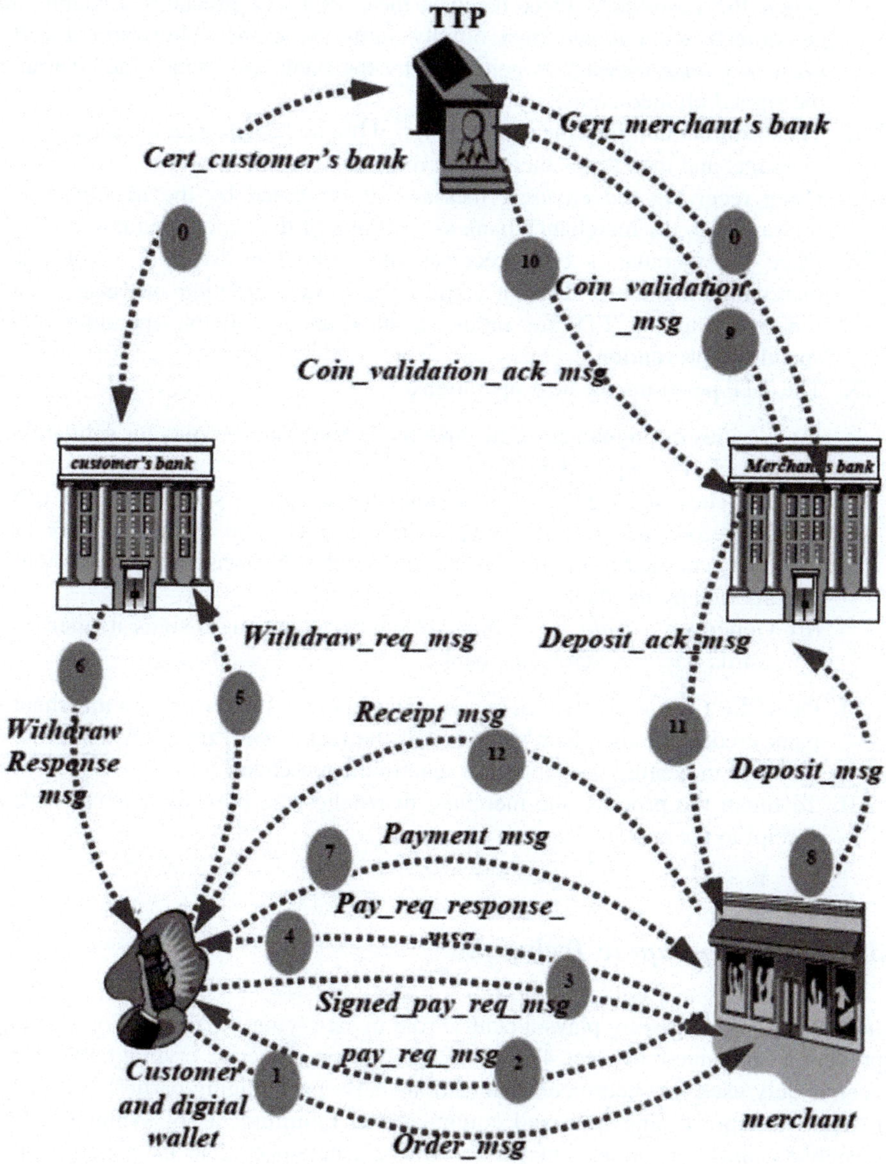


**Fig. 3.5** Overall transactions in MobiCash protocol [10]

VI. When the customer's bank receives these coins, it proceeds to debit the customer's account and then blindly signs the coins. Afterward, a *withdraw_response* message is generated by the bank, who sends the customer the signed blinded coins.
VII. The customer unblinds the received signed blinded coins, generates a *payment* message, and then sends them to the merchant.
VIII. Upon receiving the *payment* message, it is signed by the merchant and forwarded to the merchant's bank as a portion of the deposit requested.
IX. Once the merchant's bank receives the *deposit* message, it verifies the merchant's signature. Then it creates the *Coin_validation* message before sending it to the TTP for online verification and double-spending (over spending) prevention.
X. The TTP performs the next operations:
    (a) Verifies the signature of the customer's bank on the coins for validation.
    (b) Checks the expiry date of the coins.
    (c) Searches in its database to ensure that a coin is spent only one time. Therefore, a coin will be valid if it is not found in the database and then recorded in the database by the bank. Otherwise, the transaction is aborted by the bank.
    (d) Generates a *Coin validation acknowledge* message, signs it, then forwards it to the merchant's bank.
XI. Once the *Coin validation acknowledge* message is received, the merchant's bank credits the merchant's account if the result was "**true**". Afterwards, a message indicating the success is send to the merchant.
XII. To finish the process, the merchant dispatches the purchased items with a receipt to the customer.

### 3.1.2 Cryptographic Technique

In an MPS, cryptography plays a central role in satisfying the transaction security properties presented in Sect. 4. Symmetric and asymmetric cryptography have been widely used for secure communications between engaging parties, but owing to the lower processing time and computational requirements of symmetric key operations, these are more suitable for wireless networks than asymmetric ones [153, 174].

#### 3.1.2.1 Asymmetric Cryptographic

Since the first credit card appeared in the 1950s, the so-called plastic money has been widely accepted as a payment method. Traditionally, using credit cards has been done physically by presenting the credit card to a merchant and signing the

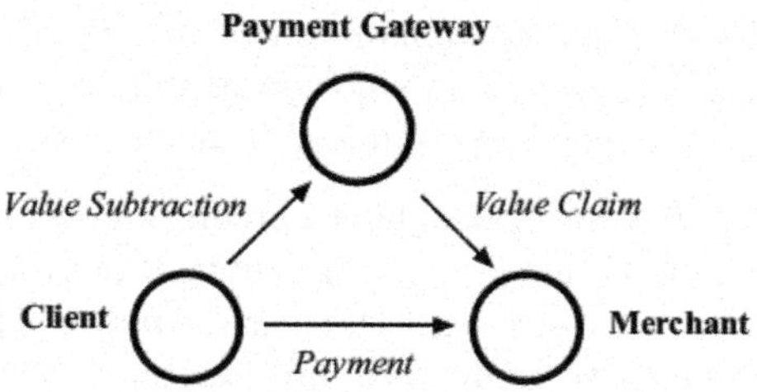


**Fig. 3.6** Primitive transactions [93]

payment slip or entering a pin number as evidence of payment. However, with the evolution of the Internet, paying with a credit card can be done online. Recently, several payment protocols have been proposed to secure credit card payments on fixed networks. However, these protocols do not apply well to wireless networks due to the limitations of wireless devices and wireless networks themselves [74, 93]. To overcome these limitations, a simple and powerful credit card payment protocol (called KSL protocol version 1 (KSLv1)), which is suitable for wireless networks, was proposed by [93]. The proposed protocol can be implemented using a secure cryptographic technique that satisfies all transaction securities, both SET [114] and *i*KP [14]. Moreover, it can resolve disputes and recover from failures that normally occur in the wireless environment. It is important to note that, in the proposed protocol, the merchant, the payment gateway, that issuer, and acquirer, need to have their own certificates.

The KSLv1 protocol, SET [114], and *i*KP [14] are based on the payment model proposed by [1] that relies on three primitive transactions (as shown in Fig. 3.6): (a) *payment,* which the client makes to the merchant; (b) *value subtraction,* which the client makes in order to request a payment gateway (on behalf of the issuer) to deduct the money from the client's account; and (c) *value claim,* which the merchant makes in order to request a payment gateway (on behalf of the acquirer) to transfer the money to the merchant's account.

The performance analysis demonstrates that the proposed KSLv1 protocol has advantages over SET [114] and *i*KP [14] in terms of transaction efficiency when applied to wireless networks. The amount of cryptographic operations for each party shows that the client in the KSLv1 protocol must perform only symmetric-cryptographic operations and hash functions, which do not require high computation, while in SET and *i*KP the client is required to perform both public key encryptions and signature verifications, which are high computational tasks. Therefore, this reduction of computational tasks at the client's wireless device confirms the advantage of KSLv1 over SET and *i*KP.

The security issues analysis of the KSLv1 protocol reveals that it satisfies the following transaction securities for any payment system [93]:

- *Party authentication* is assured symmetric encryption and a secret shared among the client and issuer (called Y). The encryption ensures that the message is generated by the client or merchant, and Y guarantees that the message has been originated by the client.
- *Transaction privacy* is ensured by symmetric encryption.
- *Transaction integrity* is guaranteed by a message authentication code (MAC).

- *Non-repudiation of transactions* is ensured by the secret Y, in that it allows to the merchant to provide non-repudiable evidence to demonstrate to other parties that the message originated from the client.

Regarding the client's credit card information, in the KSLv1 protocol, the client is not required to send credit card information (which is used only as a component for generating a set of $Y_i$ shared between the client and issuer). Therefore, it is hard to retrieve credit card information, although an attacker can intercept the message by using an efficient key generation technique.

The proposed KSLv1 protocol was mathematically analyzed to prove that it is capable of performing transactions with better performance than the SET [114] and *i*KP [14] protocols. Moreover, a design and implementation of proposed KSLv1 protocol was presented by [177] to show that it is applicable to wireless environments.

The KSLv1 implementation consists of the client-side application (the KSL Wallet) developed in the Micro Edition (JavaME) Java Platform [171]. The merchant-side and PG-side applications were written using Java Platform 2, Standard Edition (J2SE) [172].

At the client-side, the timing was measured using a Palm™Tungsten C with 64 MB of RAM and running the KSL Wallet at an IBMs Java Virtual Machine. A 2.4 Ghz Pentium® 4 processor with 512 MB of RAM was used for merchant and payment gateways. The wireless communication between the client and merchant was established using TCP-IP over the 802.11b channel WEP encryption with a 64-bit key. Screenshots of the implementation of both the merchant registration and payment phases are shown in Fig. 3.7.

In the implementation results section, the authors show that, in the first transaction, a client spent 24 s (16.84 s for the *merchant registration protocol* and 7.16 s for the *payment protocol*) on a KSLv1 payment transaction. However, because in the forthcoming transactions it is not necessary to execute the *merchant registration protocol*, a payment transaction will be completed in 7.16 s. Therefore, these results prove that the KSLv1 protocol is adequate for wireless environments [92, 93].

In e-commerce payment, the Secure Socket Layer (SSL) protocol [41] has been used to establish a secure channel between customers and merchants to secure the payment and the order information. However, SSL has some disadvantages regarding customer privacy in that the customer payment information is revealed to the merchant. In order to resolve the disadvantages of the SSL protocol, Secure Electronic Transaction (SET) [114] divided the order message into: (1) order information that is revealed to merchant M, and (2) payment information that is revealed to the Payment Gateway (PG). Even though both protocols assume the existence of Public Key Infrastructure (PKI) where extensive computation can be carried out, they are used in mobile payment, but their application is limited due to heavy computation over wireless and GSM networks [161].

In order to minimize the extensive computation of the SET protocol [114], a modified version of the SET protocol (called MSET) was proposed by [161]. The proposed protocol not only preserves the main security features of traditional

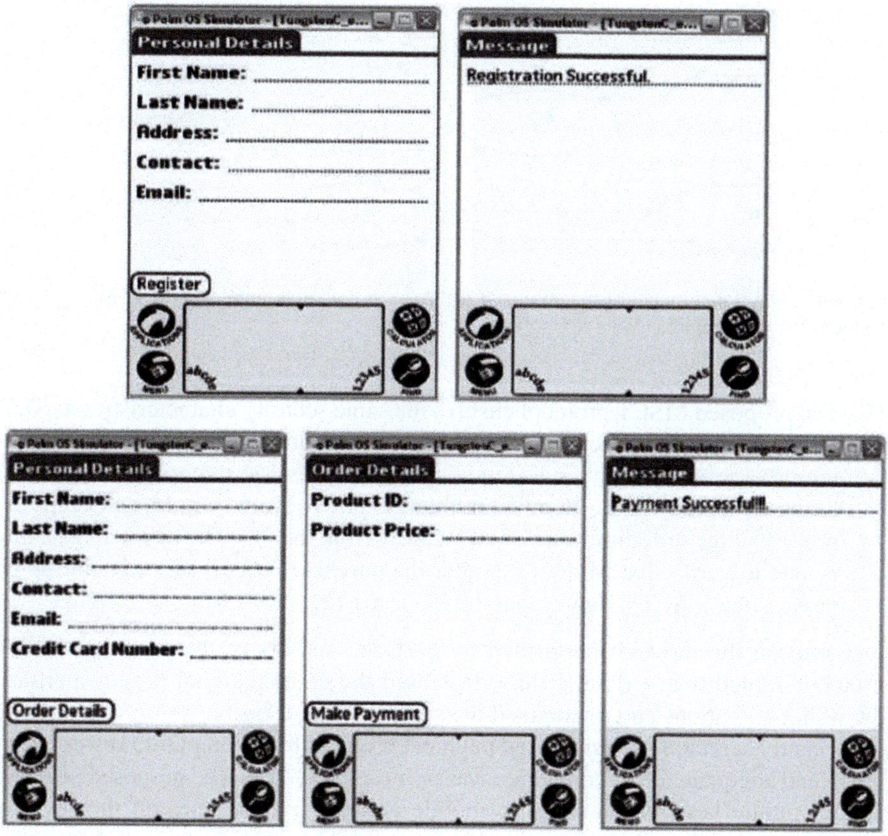


**Fig. 3.7** Snapshots of KSLv1 wallet [92]

approaches but also reduces the computational complexity of SET [114] by replacing time consuming public key encryption and decryption algorithms with symmetric key cryptography. As a result, the mobile battery power consumption is reduced due to the decreased computational complexity.

The proposed MSET protocol has the following characteristics [161]:

I. MSET protocol involves three parties as the SET protocol [114]: the mobile digital wallet (W), merchant (M), and payment gateway (PG). This makes the MSET an acceptable universal payment instrument.
II. In the set-up phase (initialization, registration), MSET uses the public key infrastructure for symmetric key exchange.
III. According to the analysis shown in [97], when comparing the efficiency of the symmetric key encryption and decryption algorithms with asymmetric key encryption and decryption algorithms, the execution speed of MSET exceeds all existing protocol at all parties.

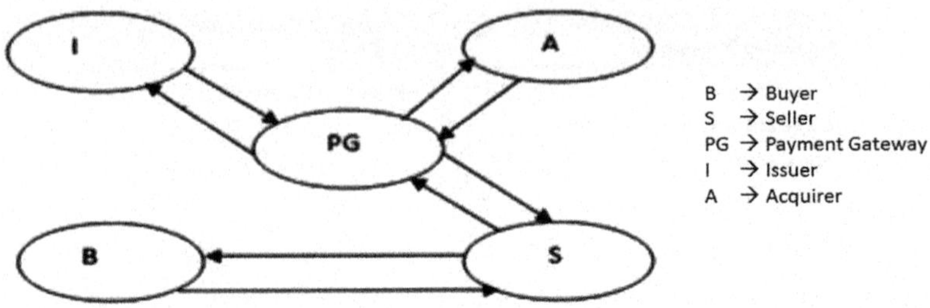


**Fig. 3.8** Operational model of the proposed secure account-based mobile payment protocol with public key cryptography [159]

IV. The proposed MSET protocol ensures the same security characteristics as SET. MSET establishes a secure channel for communication between parties, achieving privacy, authenticity, non-repudiation, and message integrity. Moreover, it keeps the privacy of the payment information of W away from M, and keeps the privacy of the order information of W away from the PG. The payment gateway is able to verify that M does not alter the purchase request message during its processing. MSET assumes that the PG is a TTP.

Based on the research performed by [97], the authors of the proposed MSET protocol argued that it dramatically decreased the computational time to perform the wallet operations and maintained the same level of security.

Recently, a secure account-based payment protocol based on public key cryptography and adequate for m-commerce was proposed by [159]. The proposed protocol applies public key cryptography to mobile network and satisfies all the security requirements of the properties provided by standard protocols for wired networks such as SET [114] and *i*KP [14]. Figure 3.8 shows the operational model for the proposed protocol, which includes five entities: buyer (B), seller (S), PG, issuer (I), and acquirer (A).

In order to analyze the performance of the proposed protocol, it is compared with two standardized protocols for e-commerce (SET [114] and *i*KP [14]) regarding the number of cryptographic operations involved for each party. Since the buyer is required to perform only one public key encryption and one decryption in the proposed protocol, the authors conclude that it has advantages over [114] and *i*KP [14] protocols. Moreover, the proposed protocol requires less computation for each party.

#### 3.1.2.2 Symmetric Cryptographic

In 2004, an account-based payment protocol (KSL protocol version 2 (KSLv2)) was proposed by [94]; this protocol represents an improved version of the KSL protocol version 1 proposed by [93] and presented in Sect. 3.1.2.1. In the KSLv2, none of the involved parties of the system is required to have a public key certificate [92].

According to the research conducted by [34], mobile payments can be classified into three different methods:

I. *Alternative payment method on Internet*, which allows users to use their phones to complete their transactions and be charged on their mobile carrier phone bill. This method is fast, gives consumers the opportunity to pay without a credit card, and does not require that the merchant invest in any special components or equipment. However, one disadvantage is that the Mobile Network Operator (MNO) imposes the restriction that only a fixed amount of money can be transferred and charged to the user's mobile phone bill.
II. *Pay at a Point Of Sale (POS) with a mobile phone* is where consumers must synchronize with the merchants system to complete a transaction. An advantage of this method is that it is useful for micro-payments. However, most of the applications require a mobile phone modification and the installation of a device in the merchant's payment system.
III. *Payment for mobile commerce applications* involves the user choosing what he/she wants to buy and conducting the transaction with a secure MPS. The main advantage of this method is that consumers can pay anytime and anywhere, but a disadvantage is that it requires mobile phone technology appropriate for mobile commerce. The third and fourth generations of mobile phones and the development of wireless technologies have helped mobile payment solutions to gain a significant market share.

In the context of the above payment methods, any person can be a potential customer who may need to purchase something. This customer could be a person who purchases from a market physically or a person who buys from e-commerce websites (called an e-buyer). For these two kinds of customers, a lightweight and secure protocol for mobile payments was proposed by [173]; this protocol provides two different payment methods. The mobile device that the customer uses in the proposed protocol plays the role of a POS device.

The proposed lightweight and secure protocol for mobile payments uses a kind of complex key to authenticate a customer and prevent his/her non-repudiation. Moreover, usage of cryptography in all data exchanges ensures the confidentiality and integrity of the protocol. Also, the employment of a complex key significantly increases customer trust because customers never enter their secure information in order to make electronic payments.

Figure 3.9 illustrates the steps performed by the proposed protocol in the first payment method where a customer goes to a market place to buy something and, after selecting her/his goods, he/she wants to pay for them. The second payment method, in which the customer is an e-buyer who tries to buy goods from e-commerce websites, is depicted in Fig. 3.10.

The security analysis of the proposed lightweight and secure protocol for mobile payments via wireless Internet in m-commerce reveals that it achieves the four major factors of the security in mobile commerce (authentication, confidentiality, integrity, and non-repudiation) using a complex key.

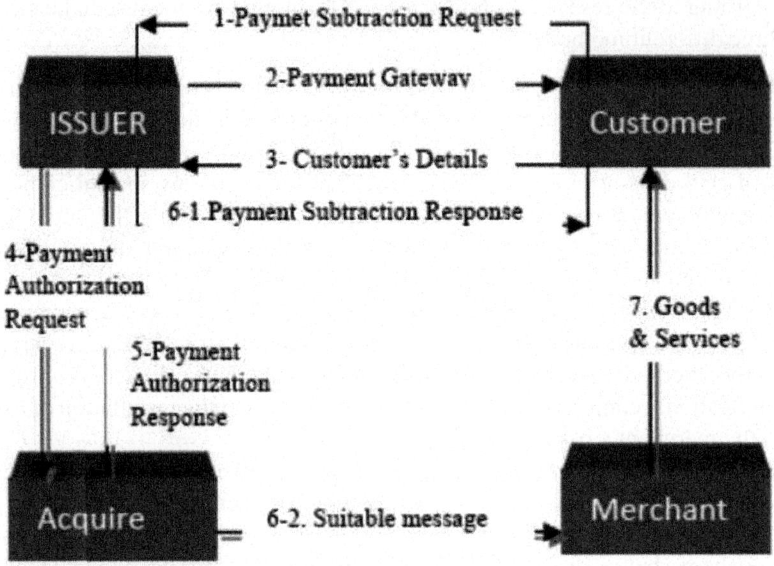


**Fig. 3.9** First method of the proposed lightweight and secure protocol for mobile payments via wireless Internet in m-commerce [173]

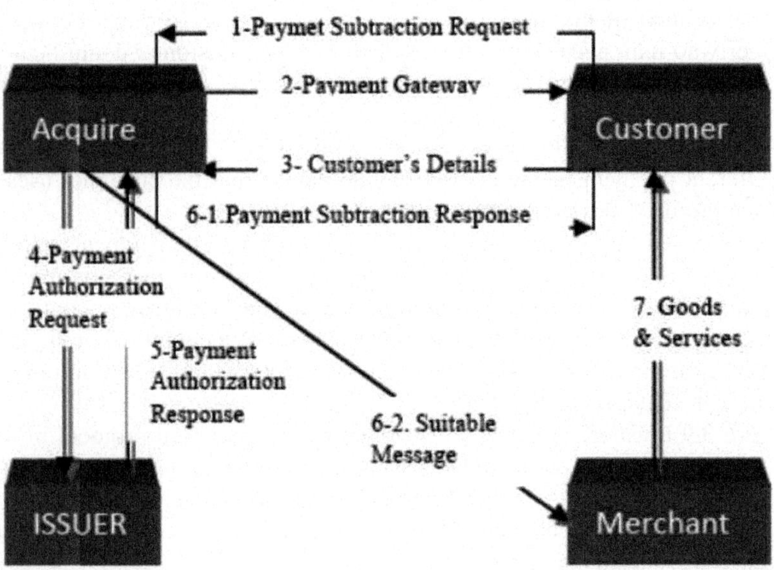


**Fig. 3.10** Second method of the proposed lightweight and secure protocol for mobile payments via wireless Internet in m-commerce [173]

An analysis of the problems that hinder the widespread acceptance of mobile payment conducted by [5] reveals that many mobile payment protocols, or Mobile Pay Center Protocols (MPPs), found in the literature are based on PKI, which is inefficiently applied to wireless networks. Others keep clients credit card information on their mobile devices or use that information in the transaction without protection, making it vulnerable to attack. Moreover, most of these payment protocols were designed to preserve, during a transaction carried out between a client and seller, the traditional flow of payment data (client—seller—seller's bank). Therefore, the data is vulnerable to attacks such as modification of the balance by the seller. Furthermore, the client's bank does not send notification to the client after the successful transfer. Additionally, design schemes of some MPCPs do not take into consideration aspects of the customer's privacy, such as customer identity.

To address the aforementioned problems, a private Mobile Payment Protocol (MPCP) which involves an MNO, employs symmetric key operations, and is based on the client-centric architecture, was proposed by [5]. The proposed mobile payment protocol minimizes the computational operations and communications between the engaging parties and ensures complete privacy protection for the payer. The proposed payment protocol is an improved version of the one proposed by [44]; the only difference is the MPCP's usage of a digital sign to avoid repudiation. However, the authors do not provide details about the digital sign generation process. The proposed MPCP consists of the seven phases depicted in Fig. 3.11.

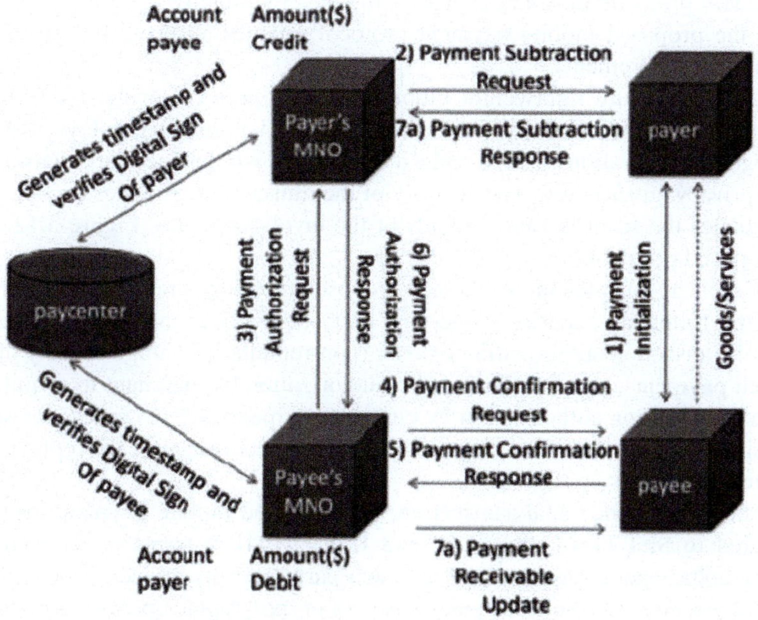


Fig. 3.11 Phases of the proposed MPCP [5]

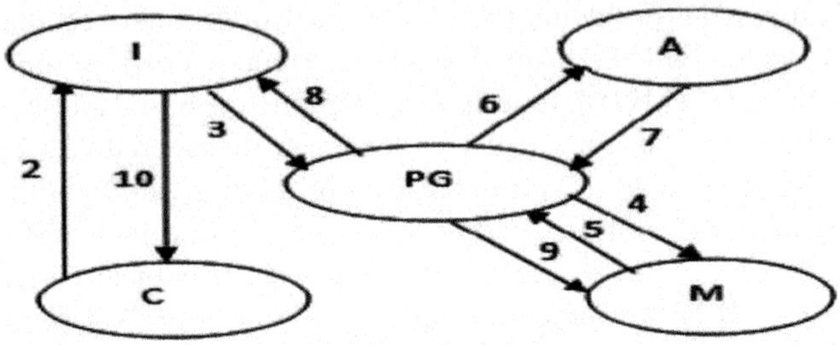


Phase 1: Payment Initialization (C ↔ M)
Phase 2: Payment Subtraction Request
Phase 3: Payment Authorization Approved Message
Phase 4: Payment Confirmation Request
Phase 5: Payment Confirmation Response
Phase 6: Payment receive Request
Phase 7: Payment receive Response
Phase 8: Payment Authorization Approved Message Reply
Phase 9 : Payment Subtraction Response
Phase 10: Payment approval Response

**Fig. 3.12** The graphical view of the proposed secure LPMP [160]

The proposed MPCP is compared to five existing payment protocols (SET [114], *i*KP [14], KSLv1 [93], KSLv2 [94], and CCMA [71]) using aspects such as privacy protection (including identity privacy protection and transaction privacy transaction) and avoids repudiation with digital signatures. The comparison analysis reveals that all the protocols provide basic privacy protection for the payer, that is, protecting the payer's identity and transaction details from eavesdroppers. However, only the CCMA protocol and MPCP protect the payer's identity from the payee. As a result, the proposed mobile payment protocol satisfies all privacy protection and repudiation requirements.

In 2012, a secure lightweight mobile payment protocol using symmetric key techniques was proposed by [160]. The proposed protocol employs symmetric cryptographic operations, which make it suitable for wireless networks, ensures customer privacy, gives end-to-end security of the transaction, provides accountability, and satisfies the security requirements of the involved parties. Figure 3.12 depicts the proposed protocol.

The security analysis shows that the proposed mobile payment protocol provides all of the following security properties: party authentication, transaction privacy, proof of transaction authorization by user (non-repudiation), impossibility of unauthorized payments, proof of transaction authorization by merchant to acquirer, and proof of transaction authorization by customer to merchant. However, this security analysis does not reveal if the proposed protocol is secure against the replay and key guessing attacks.

In the performance analysis section, the proposed mobile payment protocol is compared to SET [114], *i*KP [14], and KSLv2 [94] in terms of the number of cryptographic operations involved for each party and the privacy protection each protocol offers. As a result, the comparison of cryptographic operations reveals that the proposed mobile payment protocol has reduced the number of operations. Also, the proposed protocol satisfies more security features than the previously mentioned protocols, which is critical for the more vulnerable mobile payment scenario.

Later, in the same year, an efficient protocol for mobile payment, the Lightweight Protocol for Mobile Payment (known as LPMP), was proposed by [180]. This protocol is based on a shared secret between the client and his/her bank/financial institution and does not require any set up or registration to perform a payment transaction without any significant increase in computation time required for generating the messages handled by the protocol.

The security analysis reveals that the proposed protocol satisfies the following transaction properties: privacy, non-repudiation, integrity, and authenticity. Also, the proposed LPMP is resistant to attacks such as the man in the middle, replay, impersonation, dishonest merchant, and dishonest client.

The performance analysis compares the client-side cryptographic operations of the following protocols: SET [114], *i*KP [14], KSLv2 [94], MSET [161], MPCP [5], and LPMP. The analysis shows that the proposed protocol does not need to store, on the client device, the secret key for every merchant with which the client communicates to make payments. LPMP needs only one secret key between the customer and issuer for any number of merchants. However, the number of cryptographic operations required on the client side in the proposed payment protocol is slightly higher than the other protocols (SET, *i*KP, KSLv2, and MPCP).

## *3.1.3 Technology Used*

Technology changes rapidly and provides various possibilities for implementing MPSs. Mobile payments can use a number of different technologies to perform a transaction. The current mobile payment environment allows remote payments to rely on text messaging (SMS), a mobile browser, or a mobile app. Moreover, proximity payments rely on either bar codes or a contactless interface to chip-enabled payment technology (such as NFC-enabled mobile phones) [166].

### 3.1.3.1 Short Message Service (SMS)

SMS-based mobile payment methods can be laborious and difficult to learn, and sometimes they may not be as instantaneous as other types of mobile payment. The best instance for their use may be in long-distance communication in situations where the telephone service providers are able to provide a good guarantee of authenticity [11]. However, from the literature review made by [168], the following generic m-commerce payment schemes based on SMS were identified:

- *SMS-Based Payments:* In this scheme, secure messages are used to transfer money from one mobile user's account to another. The mobile operator gives the users a SIM card with an installed payment application and details about his/her credit card or debit card.

- *Reverse SMS Billing:* In this scheme, the mobile provider adds a charge for a special SMS called Premium SMS. The cost of that SMS would be the price for the goods purchased plus the price for sending the SMS. First, the user must dial a special service number and state the amount to be transferred along with the merchant's phone number. Then, he/she receives an SMS from the operator, which is billed accordingly.
- *Reverse Billing SMS:* This is a variation of reverse SMS billing; an additional amount is charged for receiving a certain kind of special SMS. This scheme is very popular and widely used for accessing digital content such as ring tones, music, and video, as well as special services from the mobile operator.

As the SMS is a well-established technology that has widespread use around the globe and is quick, efficient, and reliable, it has been used in recent years to design many SMS-based mobile payment systems.

A secure payment protocol for mobile devices proposed by [62] takes into account the limitations of the mobile environment in developing countries. The proposed protocol not only satisfies the convenience and ease of use that is generally required for mobile users in small payments, it also provides the transaction security level and non-repudiation property that is necessary for macro-payments. Although the proposed technique has been optimized for the current GSM network, its modular design will enable it to adapt to future improvements in mobile network technology and infrastructure, such as EMS and MMS, with minimal change in the protocol structure.

The paper [62] describes the interaction between engaging parties (including the SMS gateway) and the SMSC (Short Message Service Center); it also provides an overview of various types of payment schemes that are suitable for SMS-based payment. The transactions between engaging parties are made secure by using the AES (Advanced Encryption Standard), a symmetric cryptographic algorithm.

Also, a pilot version of the proposed payment protocol has been implemented. The simulation results presented in the paper indicate that the protocol achieves the expected level of security and can manage most security attacks while remaining cost effective and easy to use for both the users and the service providers.

In 2008, a new public key-based solution for secure SMS messaging (called SSMS) was proposed by [178]. It is an application layer protocol that simultaneously provides confidentiality, integrity, authentication, non-repudiation, public verification, and forward secrecy of message confidentiality. The basic configuration of SSMS is shown in Fig. 3.13.

Using an elliptic curve-based public key solution and symmetric encryption algorithm, the protocol achieves great efficiency, which makes it adequate for those m-payment applications where security is a big concern. Also, the proposed protocol provides the attributes of public verification and forward secrecy.

The proposed payment protocol efficiently combines encryption and digital signatures and utilizes public keys for a secure key establishment to be used for encrypting the short messages through a symmetric encryption. Besides, it has appreciable computational benefits compared with the previously proposed public

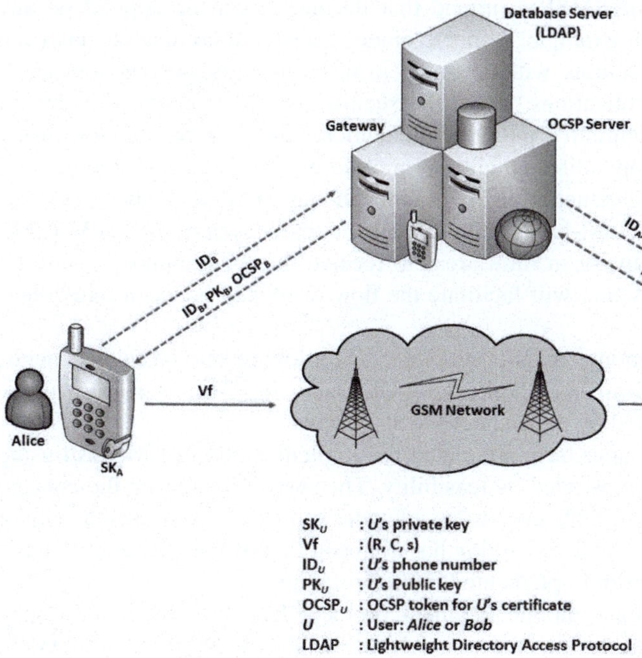


**Fig. 3.13** Basic configuration of the SSMS [178]

key solutions due to the usage of elliptic curves and a symmetric encryption algorithm. However, the system is based on public key cryptography, which requires a trusted third party to act as a certificate authority.

The SSMS payment protocol has benefits for real m-payment applications and whenever secure SMS messaging is important. It is also suitable for any other store-and-forward technology.

Another secure mobile payment model (based on SMS), called SecureSMSPay, which is suitable for macro transactions that compromise cost, simplicity, security, and the performance of the transaction with minimal cryptography key use, and fewer encryption/decryption operations compared to other models, was proposed by [61]. The proposed SecureSMSPay can use symmetric and asymmetric cryptography without the need of trusted third parties or even PKI complexity.

The SecureSMSPay is a transport channel that has the capability of sending transactions to the payer not to the payee (as is usually done in most current payment transaction models). The payer receives a secured SMS message (invoice) awaiting his/her confirmation (yes/no). Each entity (payer/payee) in the payment system trusts only his/her bank respectively, so the transaction will always go through trusted nodes. Note that the payer/payee can also use any bank payment instrument (credit card, debit card, or even current account) without revealing confidential data during the payment.

The authors of SecureSMSPay argued that the model can be applied on any payment application, for example, e-check, money transfer, e-commerce, and even normal EFTPOS transactions with leverage infrastructure supporting the above mentioned payment applications. Also, the secure mobile payment provides all security factors (confidentiality, integrity, authenticity, and non-repudiation) using a simple cryptography operation suitable for mobiles with limited resources.

Given the increasing penetration of mobile phones in rural areas, [119] was motivated to propose a GSM-based mobile money transfer system (called M-Cash) that will allow people living in rural areas to receive financial support from their relatives in towns easily and will facilitate the flow of money between individuals in such communities.

M-Cash is a mobile money transfer service ideal for low-resourced environments. It was developed in Heitml[3] and the PostgreSQL Database.[4] The interface with the GSM network is provided by the Kannel SMSC gateway.[5]

The authors of the paper have simulated the implementation of M-Cash using dummy clients and demonstrated its feasibility. The system is built on the concept of making the service available even to people who cannot afford to own a cellular phone. Note that strong SMS encryption has not been part of this phase; only basic security services have been implemented in the prototype.

Recently, a new secure mobile payment protocol based on SMS messaging (known as an SMS-based Operator-assisted Mobile Payment Protocol (SOMP)) was proposed by [95]. The proposed protocol allows a client to make a payment to a mobile operator who offers products or services or to a merchant through the mobile operator. The SOMP has the following features:

- Is lightweight as only symmetric cryptographic operations and hash functions are used.
- Satisfies necessary transaction security properties, that is, message confidentiality, message integrity, and message authentication.
- Uses a limited-use session key generation and distribution technique to prevent the reuse of session keys during the transaction.
- Is compatible with existing SMS messaging infrastructure, i.e., each SMS message allows only 160 characters (or 160 bytes).

The authors of SOMP have found that making a payment using SMS seems to be a promising application, but existing approaches to secure SMS-based payment transactions were not sufficient as they lacked both security properties and performance. The proposed payment protocol not only satisfies necessary security properties, but it is also compatible with the existing SMS infrastructure.

[3]Heitml is an XML/HTML like programming language for Web applications, which features object-oriented HTML extensions and component-based application development.

[4]PostgreSQL, often simply Postgres, is a powerful, open source object-relational database system (ORDBMS).

[5]Kannel is a compact and very powerful open source WAP and SMS gateway that is used widely across the globe for serving trillions of SMSs, WAP Push service indications, and mobile internet connectivity.

In [37], the payment protocol SOMP is formally analyzed based on the property of accountability. As a result, the authors demonstrate that the protocol satisfies necessary transaction security properties, that is, message confidentiality, message integrity, and message authentication.

M-Wallet, proposed by [145], is a mobile-based payment solution that represents an alternative for those actual payment solutions (such as credit cards, debit cards, and cash) replacing them with a simple SMS-based solution. Also, since in M-Wallet all the transactions are totally based on text messages, it is suitable for all mobile devices that support basic features like text messaging. Moreover, the proposed payment solutions is secure against spoofing of mobile numbers and replay attacks by employing in every transaction a unique one time key sent to the user's registered mobile number. A user specified PIN allows only authorized access.

The M-Wallet mobile payment is composed of the following participants: consumer, merchant, and the M-Wallet system (which is the entity liable for the performance of the payment process and the control of the flow of the transactions between the mobile consumer and the merchant). Additionally, it has the option of an M-Wallet outlet for recharge transactions.

In order to use the M-Wallet system to pay for a service or product bought from a merchant, a consumer should register by sending an SMS to the M-Wallet system using his/her mobile number, which is verified and acts as a unique ID for the user. A unique pin is generated and sent to the user, who can change the pin later for additional security. Then, M-Wallet creates a virtual account for the user with an initial balance of zero. After this step, the registered user can perform one of the transactions that take place in the M-Wallet system: recharge, withdrawal, payment, and recurring payment.

Before making payments, a user must add funds into his/her M-Wallet virtual account through an outlet of the M-Wallet system or through a bank using National Electronic Funds Transfer (NEFT). In the M-Wallet system, the recharges can be done in two ways: (a) via participating outlets, which can vary from a pharmacy to a bakery, to facilitate consumers who do not have access to banking services; and (b) via a bank transfer, which it is suitable for those who have NEFT-enabled bank accounts.

After the consumer recharges his/her virtual account, they are able to execute different transactions and pay for them using the funds in his/her account. The payment process is illustrated in Fig. 3.14. Note that the consumer can use bank transfer to redeem funds in the virtual account for money or perform a redemption at an M-Wallet outlet.

The security analysis shows that mobile number spoofing and replay attacks are prevented as every transaction is secured with a unique, single-use key sent to the mobile number registered by the user. Also, a comparison of M-Wallet with other e-payment systems demonstrates its advantages over them. However, the M-Wallet system is vulnerable to: (a) SMS delay, which can raise the time necessary to complete a transaction, and (b) loss of a message that will cause the failure of a transaction.

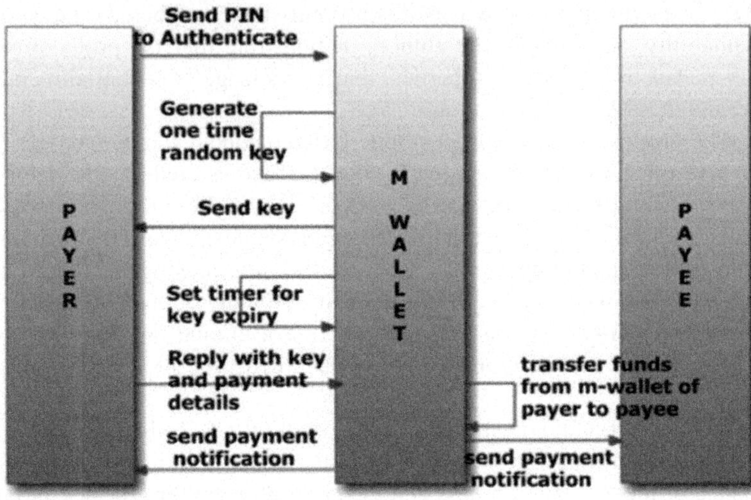


**Fig. 3.14** Payment process in M-Wallet system [145]

#### 3.1.3.2 Biometric Technology

Biometrics is a technique used to identify people by measuring some aspect of individual anatomy or physiology (such as hand geometry or fingerprints), some deeply ingrained skill, a behavioral characteristic (such as a handwritten signature), or something that is a combination (such as a voice) [91]. It is used globally and accepted for homeland protection, personnel identity verification, and social applications [109].

Biometric authentication technologies such as face, finger, hand, iris, and voice recognition are commercially available today and are already in use. The high reliability of these different biometrics technologies is derived from their randomly distributed features, which offer unbeatable protection against fraud.

For [91], a biometric system is essentially a pattern recognition system that operates by acquiring biometric data from an individual, extracting a feature set from the acquired data, and comparing this feature set against the template set in the database.

Several human traits that fit the definition of biometrics given above can be distinguished. In order to be used for recognizing a person, the human trait needs to be unique and not subject to change. However, no biometrics can guarantee 100% accuracy. The commonly used biometrics are the following: Deoxyribo Nucleic Acid (DNA), ear and face recognition, fingerprints, gait, hand and finger geometry, iris and retinal scanning, signature verification, and voice recognition.

Biometric authentication technologies are used to create a biometric payment system in which no one will have to carry dozens of cards for shopping, traveling, or opening door locks in the office, university, or bank. Compared to previous systems like credit card payment systems, wireless systems, and mobile systems, these kinds of payment systems are very safe, secure, and easy to use. The user does not even need to remember any password or secret codes [91]. Moreover, biometric payment systems are reliable, economical, and have more advantages than others.

Taking advantage of the incorporation of high performance cameras on mobile phones and PDAs, which makes possible the development of camera and image processing applications for identity verification, the authors of [109] implemented an efficient biometric mobile payment authorization system combined with an embedded commercial barcode reader to provide a worldwide platform-independent mobile payment authentication and tracking system. The objective of the proposal is to create a reliable, portable way of identifying and authenticating individuals to provide a secure mobile payment environment for both customers and merchants.

The implementation of the proposed mobile biometric authorization scheme should help to reduce greatly both customer fraud and authorization risks to acquirers (financial institutions) and mobile operators.

The pixel-oriented methods (which are extremely fast to compute and can be customized and optimized for a particular application easily) implemented in the algorithms of the proposal help to develop a low-power computing application. Moreover, via the wireless connection, the mobile can also authenticate using a large database residing on a personal computer that has a fixed IP address.

The mobile biometric identification system of the proposal is composed of four main modules: real-time image acquisition module, image segmentation and frequency analysis module, iris pattern identification module, and decision-making and authorization module. Figure 3.15 demonstrates the interactions between these modules.

#### 3.1.3.3 Radio-Frequency Identification (RFID) Technology

RFID is a technology that uses a Radio Frequency (RF) signal to exchange data between a reader and an electronic tag attached to an object for the purpose of identification and tracking. In recent years, RFID has become an important emerging technology, garnering significant interest in the wireless world and the e-commerce community by allowing systems to automatically track and transmit identification numbers using radio as the medium. There are many different applications where RFID has been used, including animal tracking, automatic toll collection, access control systems, and supply-chain management.

For [48], an RFID-enabled system consists of three parts:

I. *The RFID tag*, which identifies in a unique manner an item (or object). Every tag is composed of a microcontroller, an antenna (either made of wire or printed using conductive carbon ink), and a wrapper manufactured of polymer-encapsulating material.
II. *The reader*, which is responsible for initiating the process of identification in an automatic form by generating an RF field using a particular frequency established by the particular system. An RFID tag is discovered when it optionally authenticates the reader through a challenge—response mechanism and replies by transmitting its identifier.

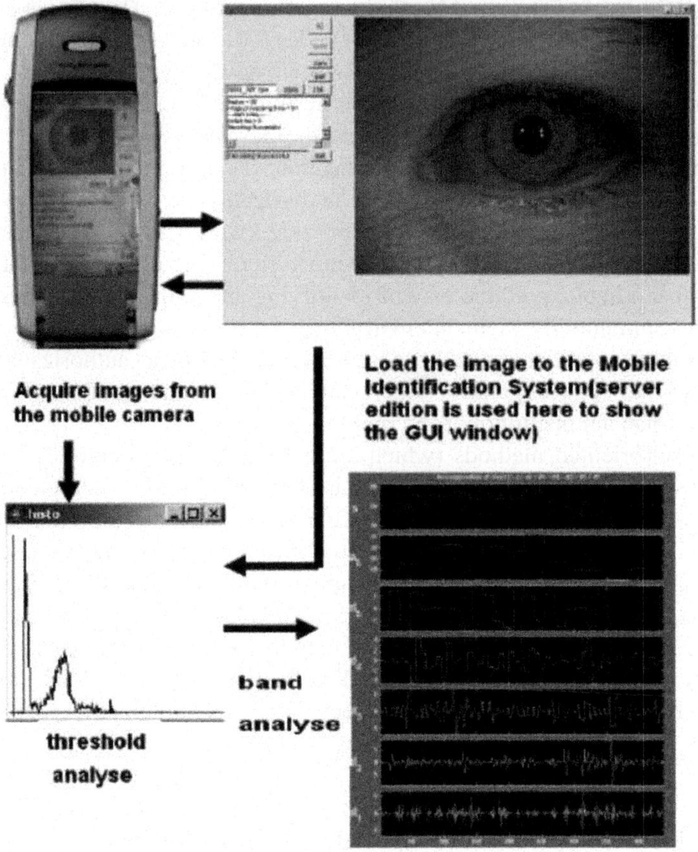


**Fig. 3.15** Mobile Biometrics Identification (MBI) system module interaction [109]

III. *The backbone supporting system*, which generally provides the pre-defined RFID tag tracking, analysis, reacting, and processing functions that allow the development of further applications.

RFID tags can be categorized into two types: (a) passive tags, which work without a power source, have a high error rate in data transmission, and usually can be read by a reader within a few centimeters; (b) active tags, which work with a power source and provide much more reliable communication in a range up to 100 m. Figure 3.16 shows some examples of RFID tags.

The rapid advances of RFID technology and m-commerce have triggered the need to develop new RFID-enabled m-commerce applications and systems that satisfy the actual and future demand despite the large number of issues and challenges in consumer privacy protection, RFID standards, and risk management [48].

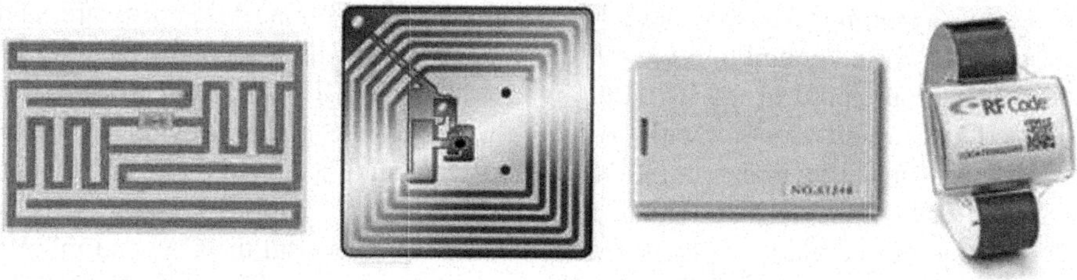

Fig. 3.16 RFID tags samples

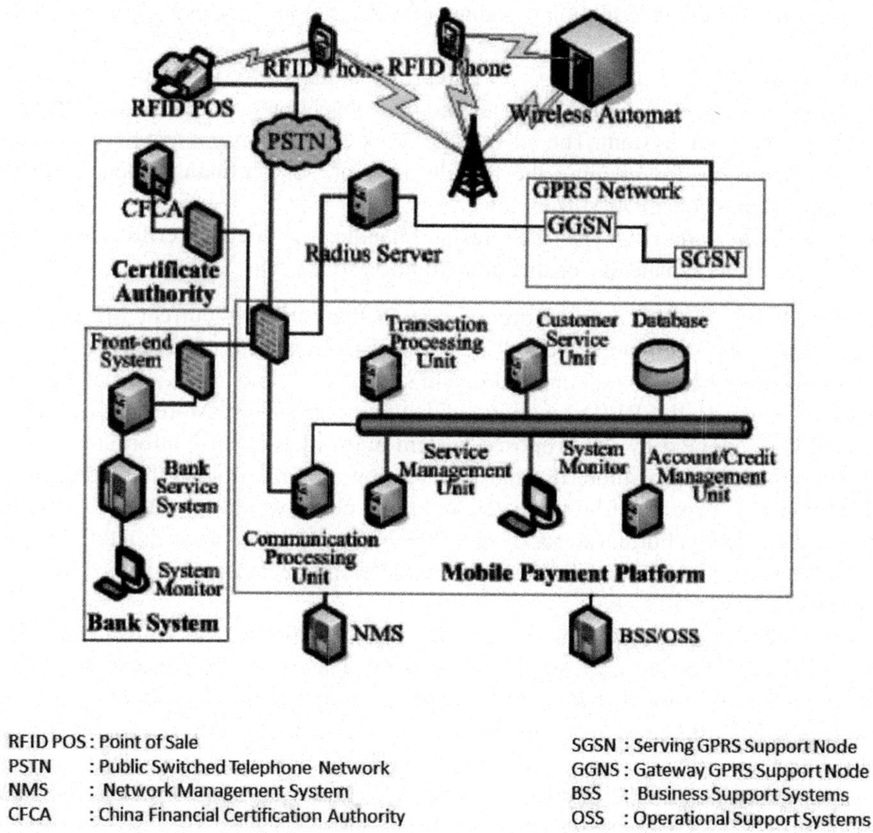


**Fig. 3.17** System structure of the GPRS MPS based on RFID [107]

After analyzing the current development of MP Service in China, a GPRS MPS based on RFID was proposed by [107]. The proposed system is more efficient, convenient, and reliable than the current Chinese MPS. Also, in this system users are not required to switch to a new mobile phone to support an RFID chip.

The proposed payment system is made up of the following parts, as shown in Fig. 3.17:

- *Mobile Terminal:* These include RFID tags, Java cell phone MPP software, and RFID POS (integrated into the R-FID reader). The RFID POS reads the clients' information stored in the RFID tag while the Java cell phone MPP software is used to swap information with the MPP through GPRS and to provide a friendly Human—machine interaction to help the user to realize the payment.
- *Communication Network:* This includes cable networks (employed principally to connect the mobile E-commerce merchants, SP, banks, and the CA) and mobile networks, which connects the mobile terminals and the MPP through GPRS and a mobile gateway.
- *Mobile Payment Platform:* This refers to the core of the MPS and provides a platform for the information exchange between mobile terminals, banks, and the CA.
- *Bank System:* This is composed of two principal parts. One is the Bank Front-end System, responsible of the protocol and the conversion of data format among the MPP and Bank System. The other is the Bank Client Management Server, which is responsible for opening the mobile account, service management, account inquiry, and consultation from the clients.
- *Certificate Authority:* This liberates and handles the clients' certification based on the China Financial Certification Authority (CFCA).

The RFID MPS is a multipurpose system that supports current MP services provided by the "Cellphone Wallet", performs usual functions of a bank card (which is the feature of MPS based on RFID), and supports off-line micro-payment and ID service owing to the RFID functions of tagging and saving. Moreover, the RFID technology adopted by the proposed system allows the clients' information to be put into the system without manual intervention (automatically); thus, the private cell phone number must be protected since the client would have to provide it to the payee directly. Furthermore, since GPRS is the service supported in this system, it is better than SMS and USSD in communication rate, interconnection, real-time online, security, and cost.

A mobile payment scheme based on RFID suitable for micro-payments and high-value payments was proposed by [43]. The proposal presents a mutual authentication protocol that uses certification and dynamic key techniques, and is secure against illegal reading, eavesdropping, camouflaging, deception, replay attack, and position tracking.

The proposed mobile payment scheme based on integrating a user's SIM card with an RFID tag can minimize the hardware cost needed for the on-site payment business based on RFID technology, and the authors claim that can be implemented easily. Compared to existing MPSs based on RFID (such as the proposal of [107]), the proposal makes real-time payments between the reader and card, saving on the cost of communication, reducing the time for payment, and avoiding some unexpected events in the payment procedure for users.

For the security of the proposed mobile RFID payment scheme, a mutual authentication protocol is presented using secure certification and dynamic key techniques. The authentication protocol focuses on the security of the data exchange between the SIM cards and the readers that is made via an RF channel. In the case of micro-payments, a trusted third party is used during the process of authentication among the vendor and the payment platform. For high-value payments, two factors are used: a user password is aggregated into the authentication process.

The security analysis of the proposed protocol shows that it overcomes several difficulties such as one-way authentication of the hash chains, the lack of forward security of the random hash locks, and the position tracking of the hash locks. Moreover, the two-factor authentication utilized in the high value payment protocol will make high value payments much more secure and reliable.

IPTV is a service that allows users to purchase through the Internet movies or TV programs that can be watched through a television. A small money payment system based on multi hash chain for IPTV service was proposed by [82], and it employs RFID-USB technology (that combines RFID and the USB key) to provide a secure and convenient environment for user identification and authentication as well as guaranteeing the anonymity of the user. The proposed protocol has a security agent among the RFID tag and the RFID-USB for FRID tags to support the plug-and-play function to make registration, withdrawal, and payment possible.

The proposed payment scheme employing RFID-USB technology uses chains of hash values. A hash key is authenticated through its digitally signed root, and the root of the hash chain is blindly signed to support anonymity. Also, the proposed payment scheme employs a prepaid account from which the payment charges are drawn and which is composed of the following protocols:

- *Registration Protocol:* In this phase, the customer, who is identified via an RFID-USB key, creates an account in the bank using arbitrary (or random) identification information ($SID_C$) and obtains from the bank their private key. Also, the merchant creates a bank account and receives his/her secret key from the bank.
- *Withdrawal Protocol:* Prior to initiating a transaction with the merchant, the customer first contacts the merchant to prepare a hash chain for payments.
- *Payment Protocol:* The payment protocol specifies the process of payment from the customer to the merchant after the customer requests the merchant's services.
- *Deposit Protocol:* In this phase, once a transaction with the customer is completed, the merchant creates a deposit request based on the customer's payment information and sends it to the bank, which verifies the deposit request sent from the merchant. If verification is successful, the bank credits the merchant's account with the amount of money to be paid.

The proposed system utilizes hash chains to implement divisible digital cash and provides user privacy protection (anonymity) by introducing arbitrary identity information. Its adopts a tracing scheme in order to protect against double-spending that occurs in the usage of hash chains. Additionally, employing of secret keys rather than public key certificates increases efficiency.

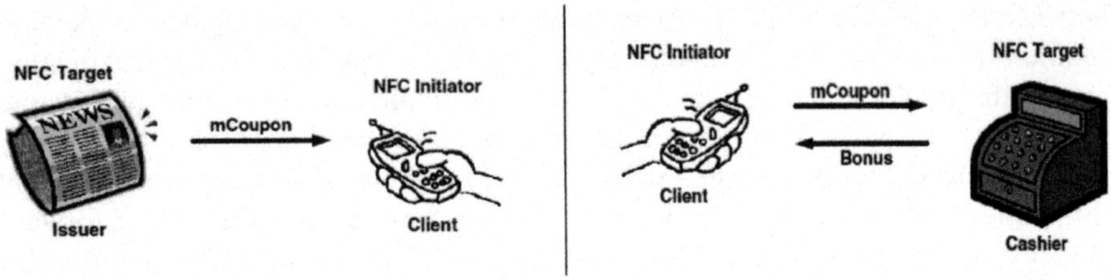


**Fig. 3.18** Basic mechanism of an mCoupon: issuing an mCoupon (*left side*) and cashing in an mCoupon (*right side*) [36]

#### 3.1.3.4 Near Field Communication (NFC)

NFC is a short-range wireless technology derived from the RFID family, which is not suitable for micro-payments. NFC has been standardized and initiated by Sony, Philips, and Nokia, who founded the NFC Forum.

Considering that nowadays coupons (paper-based or electronic) are a very common way for companies to promote their products or services, in 2007, Dominikus et al. [36] proposed a new form of coupons (called mCoupons) that are downloadable from a poster or a newspaper endowed with a passive NFC device to a mobile device that will be used by the user to cash in the mCoupon at the cashier. As a result, an mCoupon system based on NFC technology was proposed by [36]. As mCoupons are vulnerable to multiple cash-ins, unauthorized generation and manipulation, and copying, the authors proposed two protocols to address these security threats and ensure mCoupons have distinct security characteristics. The basic mCoupon mechanism is shown in Fig. 3.18.

The first proposed protocol, called the "simple" Coupon protocol, provides security against multiple cash-ins and unauthorized generation and manipulation but not against unauthorized copying. The other protocol, called the "Advanced" mCoupon protocol, employs client authentication in addition to the features of the simple protocol explained formerly (including security against unauthorized copying). The client, or better the client's mobile device, uses asymmetric authentication methods for his/her authentication and has to generate a key pair consisting of a private and a public key. Many parties that work in distinct environments take part in the mCoupons protocol. Every one of them is involved in various processes and has distinct requirements [36]:

- **Activator:** The activator shall be a device with enough computational power that is responsible for the initialization for the issuer and the cashier. The timing restrictions for initialization are also not very limited.
- **Issuer:** The issuer is a passive NFC device that takes the energy to operate from the field of active NFC device nearby the issuer. Therefore, there is an extreme limitation of energy for the AES computation. A contactless smart-card chip or even passive RFID or NFC devices equipped with a proper crypto hardware can now provide the functionality required by the issuer.

- **Client:** The client's mobile device could be a mobile phone or a PDA capable of performing asymmetric authentication and storing the private key in a secure memory. Moreover, the mobile device should have an NFC interface and software that supports the requesting, handling, and cashing-in of an mCoupon. Today, a regular mobile phone or PDA equipped with NFC technology has sufficient computational power and memory to participate in the mCoupons protocol.
- **Cashier:** The cashier is an active device with high computational power and energy resources responsible for verifying the validity of cashed-in mCoupons. It must be able to verify digital signatures and to perform an AES calculation. Using a standard PC solution with online access is possible to provide the functionality of the cashier.

In contrast to existing systems of electronic coupons, mCoupons do not need online access for the issuer and mobile client during pick-up and cash-in of coupons. Also, the requirements for each of the parties in the proposed system can be met with currently available technology.

Another proposal, which [25] proposed, is a secure MPS solution for use in a traditional in-store environment that combines the Citizen Digital Certificate (CDC) (which is a PKI-based citizen identification card given to a user by the government, under a strict user registration process), the NFC secure element within the SIM, and a 3G mobile network. Furthermore, the proposal offers a suitable user experience taking advantage of the wide-scale 3G network and the short-range contactless communication of NFC and possibly replacing the usage of payment or smart-cards.

The proposal presents the binding of NFC mobile phone security technologies with the user identity security of the CDC card that is backed by a strong user registration process. To achieve the binding, an endorsed registration phase cryptographically binds the PKI credentials of the CDC card and NFC phone in a way that is then nationally recognized. Thus, the credentials can then be used for in-store payment transactions to provide authentication, integrity, and non-repudiation without the user needing to carry any payment or ID cards. This solution is applicable to multiple MNOs and has a number of interesting features. However, the implementation performance requires further investigation [25].

In the proposed NFC m-payment system, it is assumed that a customer wishes to perform a mobile payment transaction while shopping within a conventional in-store environment (with a fixed line POS) and that the customer is already registered for CDC. The uniqueness of the CDC card, the private key and public key secure functionality, and the nation-wide acceptance and validation are complimentary features for NFC phone (SE-SIM) enabled mobile payment services. In the proposal, the m-payment transaction service is separated into the following two phases [25]:

- **Phase 1 Endorsed Registration:** This is the process of binding the mobile transactional credentials with certified customer credentials. The following three entities are used in this phase: MNO/Authentication Centre (AuC), the customer's NFC phone/SE-SIM, and the customer's ID card (e.g., CDC). In the proposed scheme, it is assumed that both the MNO and customer's CDC are under the same Certification Authority (CA), i.e., the Government CA (GCA) represents the trusted third party and is used to verify the customer's CDC so that it can be used to endorse the customer's SE-SIM. Since the MNO (which works as a domain entity to verify the mobile user's phone and associated CDC) recognizes the GCA, it can check the authenticity of the CDC the customer provides and verify the Endorsed Credential (EC) to generate a certificate for the SE-SIM ($Cer_{SE}$) for later use in mobile payment transactions.

The customer's NFC phone is a bridge for the MNO to authenticate the CDC ID card and prove that the transaction information is backed by the CDC. The main job of the CDC here is to generate the EC as a valid endorsement for the customer's phone when performing subsequent m-payment transactions.

- **Phase 2 NFC m-Payment Transaction:** Once the endorsed registration phase is successful, the customer's phone/SIM is now ready to perform in-store m-payment transactions, in which a customer tries to perform an in-store m-payment through the authentication/verification of the MNO that the CDC endorses. On first entering the payment application, the phone automatically displays the expiry date of the EC certificate to the user, and payment actions will be restricted if the EC certificate ($Cer_{EC}$) is out of date. The general payment process is that a customer holds his/her phone close to the shop NFC POS so the phone can present($Cer_{EC}$) for an ID authentication of its SE-SIM, and, if the check passes, the EC is sent in addition to the payment information.

The proposed protocol was considered with respect to a number of attack scenarios, demonstrating that it is resistant to those scenarios. Moreover, the solution, which is applicable to multiple MNOs, has a number of interesting features, although the implementation performance requires further investigation.

Recently, Mainetti et al. [111] presented an innovative and secure NFC micro-payment system based on the peer-to-peer NFC operating mode for Android mobile phones (called IDA-Pay and developed in the IDA-Lab (IDentication Automation Laboratory) of the University of Salento, Italy), which can be used easily in alternative micro-payments ecosystems. The proposed system should ensure a security level equal to that of conventional credit card payments without the participation of any hardware (SIM or SD cards replacement) by the owners of smartphones. This means that the application must be downloaded by the users from the market and installed onto the smartphone [111].

In order to validate the proposed IDA-Pay system, an application prototype (called the IDA-Pay app) for the client side, was implemented and executed by using a Samsung Nexus S mounting Android 4.1.1 (i.e., Jelly Bean version). The package javax.crypto from the Android SDK has been used in order to produce the Credit Card File (CCF). On the other hand, the IDA-Pay POS prototype uses a SCM Microsystems SCL3711 USB Dongle NFC reader.

The validation scenario is illustrated in Fig. 3.19 (where it is assumed that the buyer has configured at least one credit card into his/her smartphone) and is described as follows [111]:

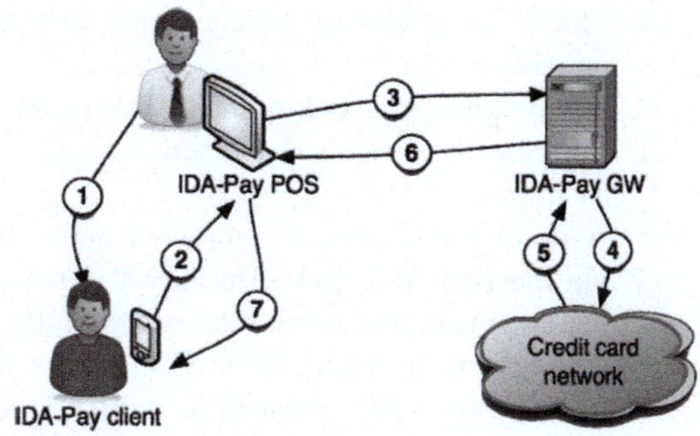


**Fig. 3.19** IDA-Pay validation scenario [111]

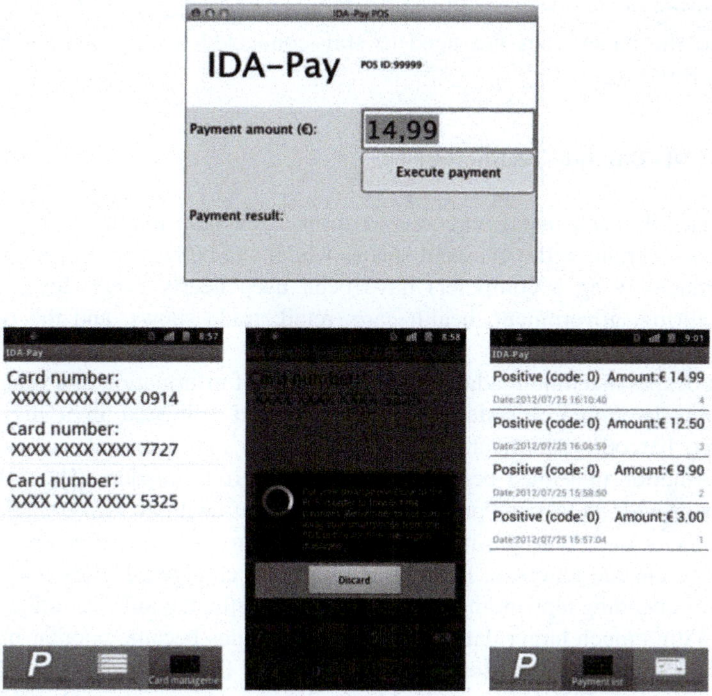


**Fig. 3.20** The IDA-Pay prototype [111]

I. The cashier inserts and confirms the amount into the IDA-Pay POS application (see Fig. 3.20, top) and asks the buyer to touch his/her NFC Android smartphone to the POS NFC reader.
II. The buyer launches the IDA-Pay app and selects a credit card from the configured credit card list (see Fig. 3.20, bottom-left). The application prompts the buyer to insert the corresponding PIN and then ask the buyer to touch the smartphone to the POS NFC reader (see Fig. 3.20, bottom-center).
III. Payment details are transferred to the IDA-Pay Gateway (GW).

IV. The IDA-Pay GW instantiates a payment transaction with the right credit card network.
V. The result is received by the GW from the credit card network.
VI. The result is formatted and sent back to the IDA-Pay POS to be shown to the cashier.
VII. A plain text response is created and sent back to the buyer's smartphone via the NFC P2P link. The buyer is notified, through a process dialog, that the payment has been sent successfully and that he/she can separate the smartphone from the POS NFC reader. The payment response is stored in the payments archive (see Fig. 3.20, bottom-right).

Note that the buyer's smartphone must remain in touch with the POS NFC reader during the entire transaction in order to receive a response from the system. However, the buyer does not need to stay connected to the Internet during the payment process.

#### 3.1.3.5 2-D Barcode Technology

In the last 30 years, linear barcodes (a form to encode numbers and letters in a sequence of variant width bars and spaces which can be read, retrieved, processed, and validated using a computer) have been used nearly everywhere, including manufacturing, government, health care, retail, trade shows, and the postal and automotive businesses.

Barcodes, as machine-readable representations of information in a visual format, can be easily stored, transferred, processed, and validated in a digital form. Moreover, barcode identification provides an easy and economic way to encode text information that must be readable using electronic readers. Thus, the use of barcodes provides a fast and precise tool to enter data without keyboard data entry.

In order to overcome the restriction of the forms of linear barcodes about the capacity to encode letters, at the end of the 1980s emerged 2-D barcodes that are capable of encoding alphanumeric data, including letters, numbers, and punctuation marks. With a much larger data capacity, 2D barcodes became popular in different areas. Two kinds of 2D barcodes can be identified (see Fig. 3.21): (a) stacked 2D barcodes (such as PDF417) and (b) Matrix 2D barcodes (such as Data Matrix and QR Code) [47].

QR Code PDF417 Data Matrix

**Fig. 3.21** 2D barcode samples [47]

As mobile data entry represents one critical issue in the development of m-commerce systems, the use of barcodes enables this issue to be solved in a simple and effective way owing to advantages over linear barcodes in data capacity and visual representation size. Therefore, the use of 2-D barcodes will enhance the mobile payment experience of mobile users.

In 2009, [47] proposed an innovative MPS based on 2D barcodes to improve mobile users' experience in conducting mobile payment transactions. It allows mobile users to buy and sell as well as pick-up goods identified with 2D barcodes anytime and anywhere. The proposed MPS was developed using the same approach to building 2D barcodes in MPSs, enabling mobile users to emit m-payment transactions using their digital wallets based on mobile payment accounts in a mobile payment server.

This approach offers advantages over existing account-based MPSs. It enables mobile users to retrieve all the relevant information of products from 2D barcodes in a easy manner, increases the mobile security of payment transactions, and enhances the experience of mobile users through the reduction of users inputs.

The infrastructure for the proposed system is shown in Fig. 3.22, where the following layers are evident [47]:

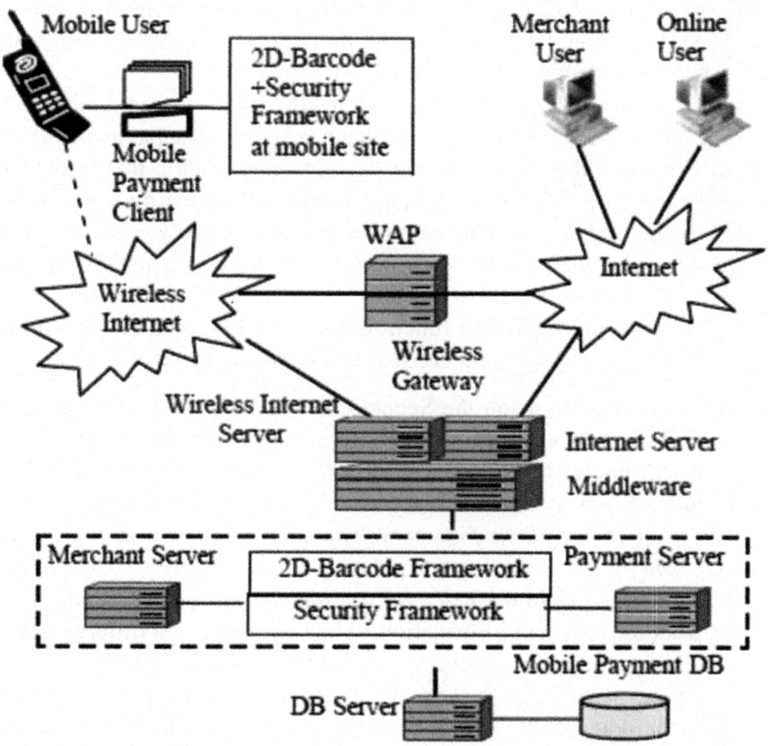


**Fig. 3.22** The 2D barcode-based payment system infrastructure [47]

- **Client Layer:** Two interfaces are included in this layer: (a) the online client interface, designed to support end customers and merchants by performing user account and membership management, and (b) the mobile payment client software (mobile user interface) that supports mobile user interactions for mobile payment. The mobile payment client is supported by the Java Platform Micro Edition (JavaME) [171], Bouncy Castle Light Weight Crypto APIs [175], JSON utility [84], and 2D barcode framework [49].
- **Middle Layer:** This layer is composed of the following components: a wireless internet server (Tomcat server) and other Java related middleware (such as Java Platform 2, Standard Edition (J2SE) [172], Bouncy Castle Crypto Library, JSON utility, Java DB connectivity, and Java Servlet technology [193]).
- **Application Layer:** This represents the mobile payment server responsible of the 2D barcode-based mobile payment process and related transactions.
- **Data Store Layer:** This layer consists of a database server responsible for storing, maintaining, and processing the customer account and membership data, transaction data, the digital wallet of the user, and the required security data such as session records and PIN.

In order to demonstrate the performance and benchmarks of the proposed 2D barcode-based payment, a prototype was build to test it on a wireless Internet connection using a mobile emulator. The result presented in the paper shows that the total processing time taken for completing a mobile payment transaction using a server-side encode/decode algorithm was 43,469 ms.

Recently, a Secure QR-Pay system based on QR-code was proposed by [99]. The proposed system avoids using a secure channel to transmit payment information by using a QR-Code that expresses the payment information that is displayed in a window by the shop. Thus, a user can catch the QR-Code by using the camera attached to the mobile device. This payment system is very easy to use and secure.

In order to utilize the proposed Secure QR-Pay System, a user needs a mobile device with a camera and an Internet connection, and the shop requires a terminal with a display and communication functions (such as a desktop PC with an Internet connection). The proposed system consists of the following steps [99]:

I. *Initialization:* Before using the Secure QR-Pay System, a user and shop must obtain their certificates (which includes a public key and a private key that only owners know) through the Public CA. Then, they have to register via the Payment Gateway (PG). Afterwards, the users must install on their mobile devices the *QR-Pay client for users*. The shop must also install the *QR-Pay client for shops*.

II. *Shop's Payment Information Creation:* When a user buys goods at a shop, he/she asks for payment information (which consists of a differentiating shop number, unique payment number for each transaction at the shop, and details of both payment due and the total payment made). When the purchase information has been introduced, the shop inputs the payment information. The payment information and the shop's digital signature are transmitted to the PG. Then, the PG requests the public certificate for the transmitted shop numbers from the CA to verify payment information. The PG extracts the public key from the shop certificate and verifies the digital signature of the transmitted payment information. If the verification of the digital signature is successful, the PG registers the payment information as "Request" on the payment information list page and knows about it at the shop. The shop shows the shop number, payment number, and digital signature to users in the QR-Code format.

III. *User's Payment Information Confirmation:* Using the camera attached to the mobile device and the QR-Pay client for users program, a user can take a photo of the QR-Code created in the previous step. The QR-Code contains the shop number, payment number, and digital signature value. Using the shop number and payment number, the client program can download the payment information related to payment matters on the payment information list page of the PG. To verify the transmitted payment information, the user must request the shop's public certificate from the CA to extract his/her public key. Then, the digital signature can be verified using the extracted public key. If the digital signature is verified, the user's terminal shows the details of the payment information.

IV. *User's Payment Approval:* In this step, the user checks the payment information shown on the terminal during step three and the details of the user's purchase in order to approve the payment. For the final approval of the payment, the user signs digitally using the private key that is stored on their mobile device and protected by a password. After verifying the password, the client program signs to pay the total amount in the payment information by using the private key. Then, the shop number, the payment number, and the payment approval are transmitted to the PG. The PG then verifies the received payment approval using the public certificate from the CA. When the verification is complete, the PG changes "Request" into "Accept".

V. *Payment approval confirmation:* As a user's payment approval is completed, PG sends to the shop the reporting message completed and the information payment associated. The PG changes "Accept" to "Finish" once the payment is completed and has not been canceled or changed. The change of payment information to "Finish" confirms the completion of the payment approval procedures.

The design and implementation of the Secure QR-Pay system is presented by the authors in the research. Figure 3.23 shows the Secure QR-Pay system architecture, which is explained as follows:

- **Payment Gateway(PG)**, which is composed of the Web Server and the Verify Processor. The *Web Server* offers the payment information list page for user confirmation and supports the SSL/TLS protocol while the *Verify Processor* verifies the digital signature during the payment information verification.
- **User**, which contains the *PG Connect Processor*, the *QR-Code Decode Processor*, and the *Digital Signature Processor*. The PG Connect Processor can send and receive data received from the PG securely using the SSL/TLS protocol. The QR-Code Decode Processor converts into values the data contained in the photo of the QR-Code. The Digital Signature Processor verifies the digital signature with a certificate of authentication.

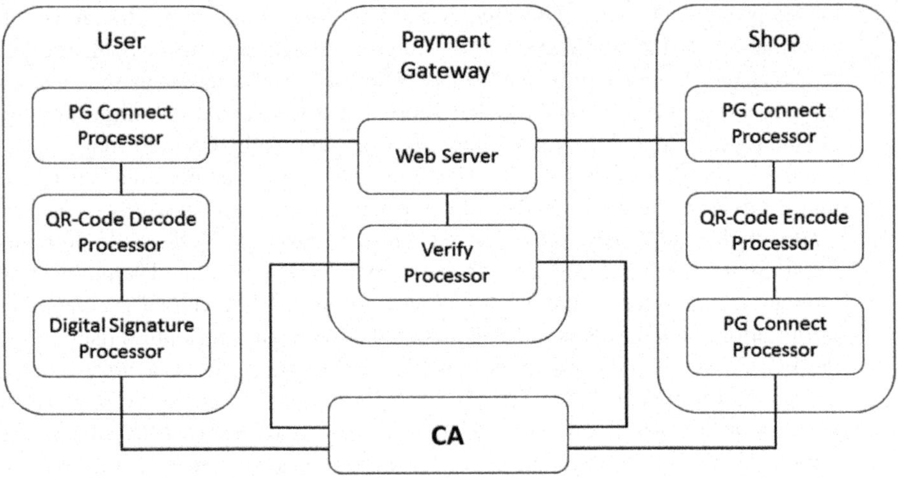


**Fig. 3.23** Secure QR-Pay system architecture [99]

- **Shop**, which is composed of the PG Connect Processor and the QR-Code Encode Processor. The PG Connect Processor works in the same manner as in the User component. The QR-Code Encode Processor encodes the payment information into a QR-Code and shows it on the monitor of the terminal.
- **Certificate Authority(CA)**, which is the Public Certificate Authority that stores all digital signature certificates of the system users.

The authors do not give details regarding the implementation of the Secure QR-Pay system and only show three snapshots of the implemented system (see Figs. 3.24, 3.25, and 3.26.).

The security analysis of the proposed QR-Pay system reveals that it is resistant to replay and relay attacks. Moreover, it provides mutual authentication, non-repudiation, confidentiality, and integrity using the public certificate of the CA.

#### 3.1.3.6 Peer-to-Peer (P2P) Technology

Due to the increased relevance of P2P in a wide variety of areas (such as distributed and collaborative computing with special regard to the Internet) in the last decade, industry and academia have carried out research and development of new activities and protocols in this area. There are many different views of how to best describe P2P technology (see Fig. 3.27). In general, peer-to-peer is about sharing: giving to and obtaining from the peer community. Also, the term P2P refers to a class of systems and applications that employ distributed resources to perform a function in a decentralized manner. Some of the benefits of a P2P approach include improving scalability by avoiding dependency on centralized points, eliminating the need for costly infrastructure by enabling direct communication among clients, and enabling resource aggregation [106, 118].

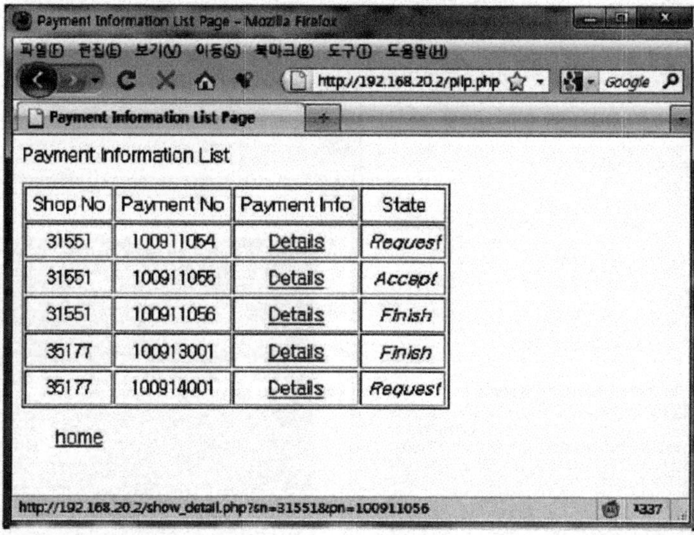


**Fig. 3.24** QR-Pay server for PG [99]

**Fig. 3.25** QR-Pay client for user [99]

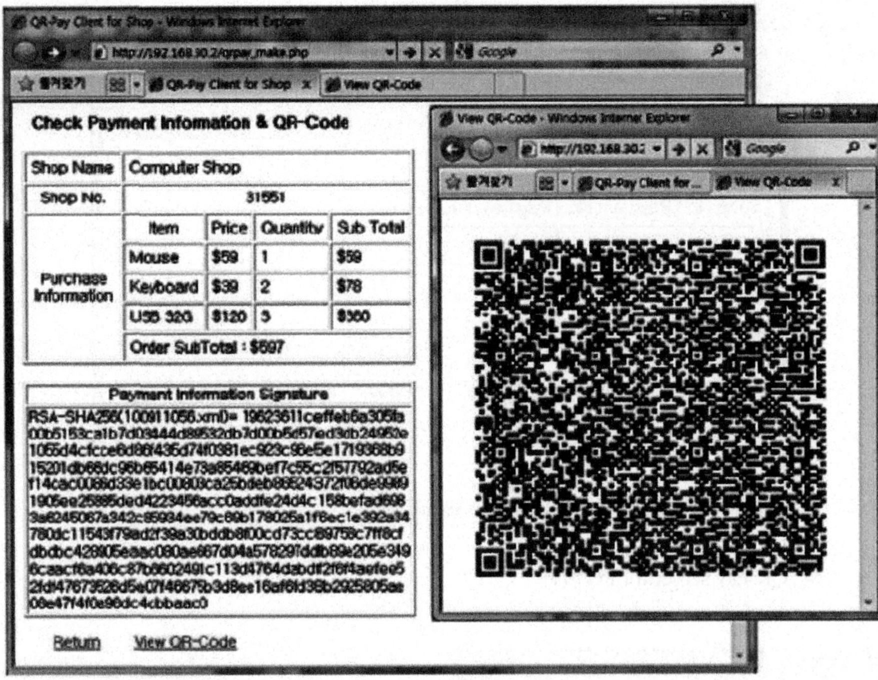


Fig. 3.26 QR-Pay client for shop [99]

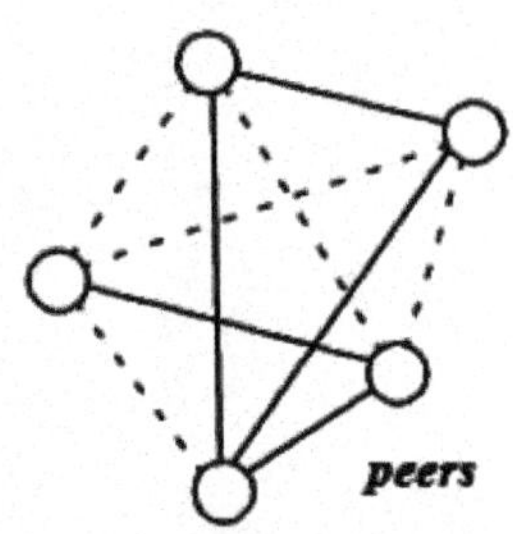


Fig. 3.27 High-level view of P2P approach [118]

In 2005, [46] proposed a P2P wireless payment system (called P2P-Paid) to permit two mobile users to execute wireless payment transactions over Bluetooth communications. The proposed system utilizes a 2D secured protocol to support the P2P payment transactions between two mobile clients using Bluetooth communications and the related secured transactions between the payment server and mobile clients.

The proposed system offers the following significant functions to mobile phone users [46]: (a) *P2P party discovery*, where both the payee and payer discover each other over the Bluetooth network; (b) *P2P session management* to allow a user to establish or drop a P2P payment session over the Bluetooth network; (c) *P2P payment management*, which enables a user to conduct a P2P mobile payment transaction over the Bluetooth network based on an established P2P payment session; (d) *account management*, which permits a user to execute basic

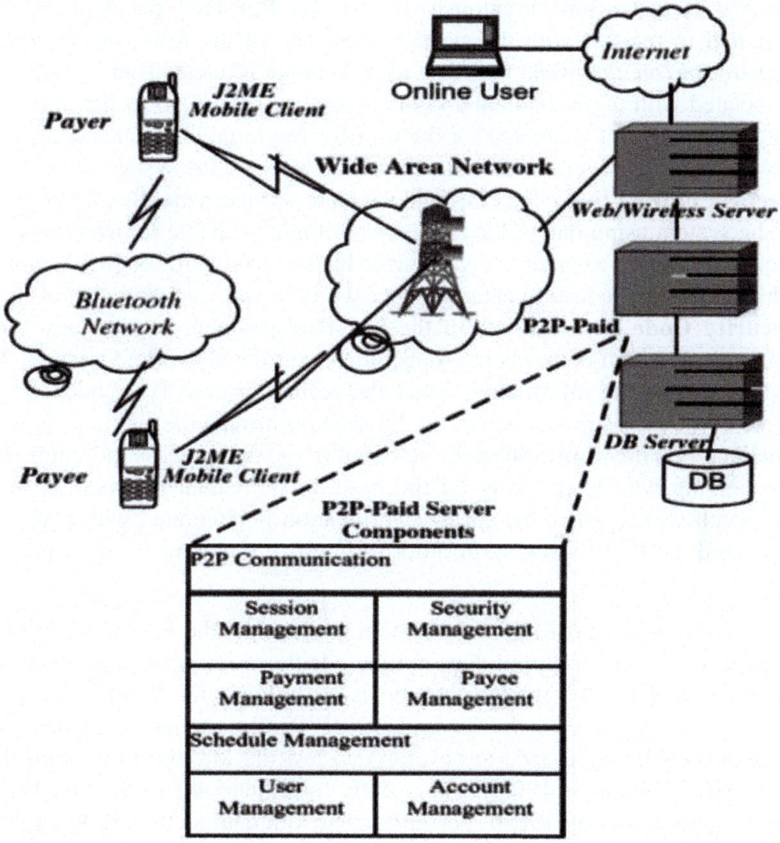


**Fig. 3.28** P2P-Paid system architecture [46]

account management functions like checking account balances, and so on; (e) *payment scheduling*, which allows a user to set up, update, delete, and view payment schedules using mobile phones; and (f) *payee management*, which permits a user to execute payee management functions, such as adding, editing, deleting, and viewing the payee information, on a mobile phone.

Figure 3.28 shows the 4-tier architecture of the P2P-Paid system, which includes: (a) *mobile lient*, which provides the J2ME-based P2P mobile client software for mobile phone users and the HTML-based online client for online access; (b) *middleware*, including an Apache Tomcat Web Server with wireless Internet support and other middleware software (such as Java JSP and Servlets); (c) *payment database server*, which refers to the database server that works with the database access programs to maintain necessary mobile user and account information and mobile payment transactions; and (d) *P2P-Paid server*, which works with middleware by communicating with the mobile client software to support the wireless payment functions. The P2P-Paid system provides the following security features:

- **Service Registration:** Previously to use the P2P-Paid payment services, a user had to register with the system using one of the following methods: (a) *registration on the Web*, where a user account is created and a key pair is associated with the web client and the system server, or (b) *mobile registration*, where a voiceprint is created for the user for future mobile authentication and a key pair is also associated with the mobile device and the system server.
- **Access Control:** To use the P2P-Paid payment service, a mobile user has to log in to the system using one of the two types of login available (a) *web login*, which requires the user to enter the valid user ID and password, or (b) *mobile login*, which requires the user to enter the user ID, PIN, and voiceprint.
- **Security Code Attachment:** In the P2P-Paid payment service, any sensitive data are encrypted before being transmitted over the network. A security header (which carries the information about the secured protocol) is attached to each piece of the message sent across the Internet or through the air.
- **Speaker Verification:** Speaker verification is the process of automatically recognizing who is speaking on the basis of individual information included in speech waves. Also, the speaker verification is integrated with conventional password (or PIN) checking in order to improve the authentication procedures [46].

The first version of the P2P-Paid system implements the P2P payment protocol with basic security support but does not include the voice verification feature. The technology used for the implementation is as follows: an Apache Tomcat Web Server (version 5.0.28) for the server, a J2ME wireless toolkit emulator (version 2.2) on the mobile client side to develop and test the Mobile Information Device Profile (MIDP) client, a JSR 82 (java APIs for Bluetooth) to test the Bluetooth communications, and the kXML parser to parse xml data on the MIDP client and a Xercer parser for the P2P-Paid server.

Some years later, the authors of [31] proposed an adaption to a P2P-NetPay protocol based on the client-side e-wallet NetPay protocol [23] that is suitable for P2P-based network environments. The new model, P2P-NetPay, is a micro-payment protocol characterized by off-line processing that uses lightweight hashing-based encryption and is suitable for P2P network service charging. Furthermore, it uses touchstones that are signed by the Central Indexing Server (CIS), which is the broker in NetPay protocol, and an e-coin index signed by peer vendors.

The proposed P2P-NetPay micro-payment system includes peer users (e.g., downloading files), peer vendors (e.g., file hosts), and a broker (which is assumed to be honest and is trusted by both peer users and peer vendors). The micro-payment only involves peer users, peer vendors, and a broker who is responsible for registering peers, crediting peer vendors' accounts, and debiting the peer users' accounts. Figure 3.29 illustrates key P2P-NetPay component interactions.

The proposed P2P-NetPay protocol satisfies the security requirements that a micro-payment system should have since it prevents peers from double-spending and any internal and external adversaries from forging. Moreover, it is efficient since it only involves a limited number of public key hashing operations per purchase.

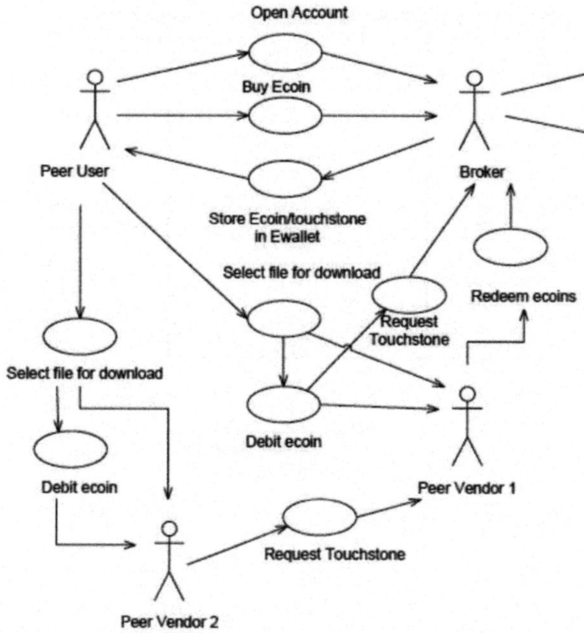


**Fig. 3.29** Basic P2P-Netpay component interaction [23]

At the end of 2008, Satoshi Nakamoto (a pseudonym) proposed a decentralized cryptographic currency system called Bitcoin [124]. The system is based on a P2P architecture and relies on digital signatures to prove ownership and provide a public history of transactions (shared using a P2P network) to prevent double-spending [147]. Bitcoin is a payment system that was released as open-source software in 2009 where users can transact directly without needing an intermediary.

### *3.1.4 Session Initiation Protocol (SIP)*

The SIP is an application-layer control protocol that can establish, modify, and terminate multimedia sessions (conferences) such as voice and video calls over the Internet Protocol (IP) [151]. Some feasible application examples of SIP include video conferencing, streaming multimedia distribution, instant messaging, presence information, file transferring, and online gaming.

Instant Messaging (IM) is defined as the exchange of content between a set of participants in near real time. These messages are usually short text messages, although not necessarily. Generally, the messages that are exchanged are not stored, but this also need not be the case. IM differs from email in common usage in that instant messages are usually grouped together into brief live conversations, consisting of numerous small messages sent back and forth [20].

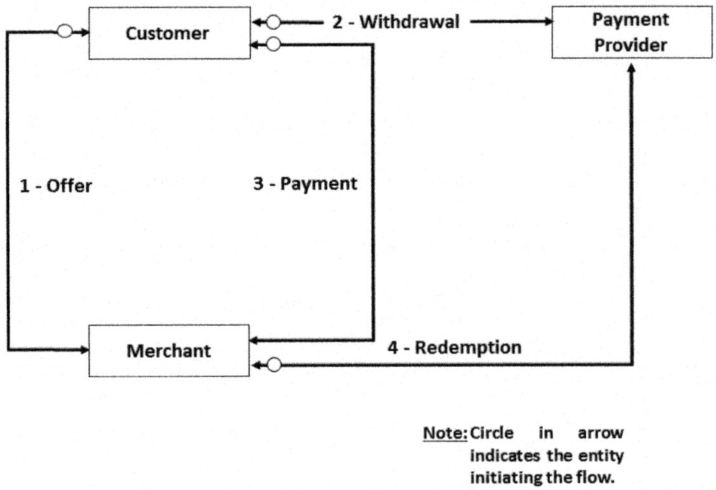


**Fig. 3.30** Basic framework of SIPCoin [60]

Analyzing the current payment approaches for SIP services reveals that those schemes are too complex to be used in real time since they have an important impact on SIP services. Therefore, aiming to enhance these schemes, [60] proposed SIPCoin, a P2P real-time payment scheme for SIP services based on hash chains. It offers real-time payment for SIP services by integrating a hash chain-based micro-payment mechanism with the SIP session establishment and management process. Moreover, it was designed to support several billing models and multiple currencies, and could be adapted to modified the number of simultaneous payment system users.

The basic framework of SIPCoin is depicted in Fig. 3.30; the entities and the trust relationship between those entities are the same as SIP Payment [81]. As the payment provider maintains the accounts of the customer and the merchant, some mechanism is needed to allow these entities to transfer money safely to and from their accounts. However, the mechanism to fulfill this functionality is not explained in the work. Therefore, it is presumed that the customer and the merchant have completed the aforementioned steps prior to the payment flow.

The basic security of the proposed scheme is provided by the irreversibility and non-repudiation of the hash chain, which hinders the customer, merchant, and others from forging SIPCoins. Moreover, SIPCoin allows users to select the current SIP security mechanisms due to its flexible interface. This could offer to users a certain trade-off between efficiency and security in accordance with the requirements of the real-world payment systems. Also, SIPCoin is resistant to the following attacks [60]:

- *Thievery Payment Attack:* An adversary tries to obtain the SIPCoins by eaves-dropping on the communication between the customer and the payment provider or between the customer and the merchant.

- *Replay Attack:* An attacker eavesdrops on messages that include significant payment information in order to repeat them to receive SIP services in an illegal manner.
- *Man-in-the-middle Attack*: An intermediary (a SIP proxy) between the customer and the merchant steals the SIPCoins in transmitted messages to impersonate the customer in order to interact with the merchant in forthcoming sessions.

From the performance analysis the authors of SIPCoin made, it can be seen that the proposed payment scheme has higher efficiency, better real-time performance, and greater extensibility when compared to other payment schemes for SIP services.

The rapid advancement of 3G technologies in recent years has added additional benefits to the next generation mobile phone network. Due to its bandwidth, the 3G network is a desired infrastructure through which new mobile payment schemes are unfolded. Accordingly, [195] proposed a SIP-based Mobile Payment Architecture (SIMPA) for next generation mobile networks. The proposed scheme supports various classic Internet security protocols to improve privacy, confidentiality, and integrity during transactions. Also, the proposed scheme supporst P2P payment communications among customers and merchants using the SIP.

The SIMPA solution employs the IM extension of the SIP protocol to carry payment information, resulting in a scalable, compatible, and secure payment platform. The requirements of compatibility, security, and low payment costs have promoted the use of SIP-based IM as the method for carrying payment information in SIMPA.

The topology of the proposed SIMPA mobile payment model involves the following six separated entities (see Fig. 3.31 [195]):

- **Customer:** Refers to the participant who pays for goods or services via his/her mobile phone.
- **Merchant:** Refers to the participant who provides goods or services and accepts the customer's payment.
- **Lock-Keeper:** Refers to a device based on the principle of "physical separation" that works like a sluice to provide secure data exchange among the physically divided networks.
- **PG:** Refers to a trusted third party that communicates with the customer and merchant to realize the transaction.
- **SIP servers:** Refers to the participant that regularly routes all the SIP messages (which contain payment information) to or from the PG. Also, in case of impersonation, it is responsible for the customer's authentication.
- **Accounting Database:** Refers to a database server that records user accounting information.

SIMPA's payment protocol is derived from Secure Electronic Transaction (SET) [114] as it can accomplish privacy protection by separating Order Information (OI) and Payment Information (PI). Thus, a customer can send a purchase message to the merchant that includes OI to confirm the purchase of items and PI that holds payment details (e.g., PIN). It is important to note that the merchant does not need

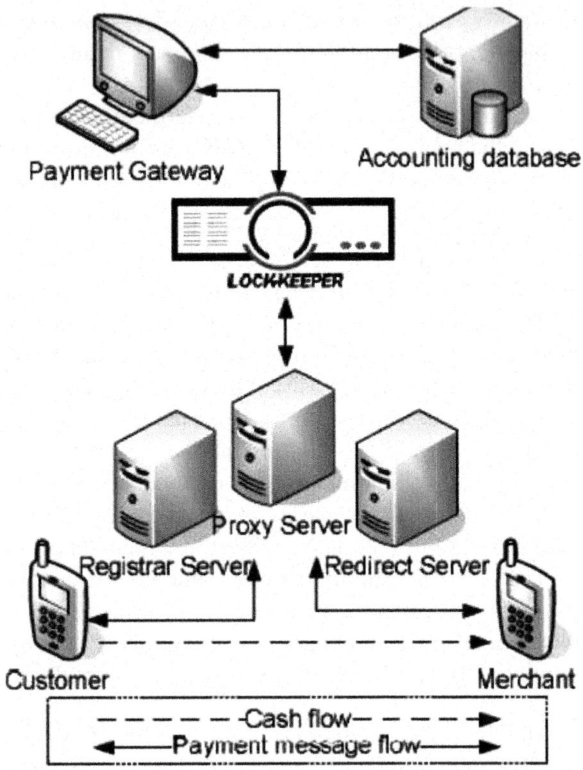


**Fig. 3.31** The topology of SIMPA [195]

to know the customer's PIN included in the PI, and the PG does not need to know which item the customer wants to buy. Once the request is received, it is decrypted by the merchant to extract the PI, which will be forwarded to the PG to complete the transaction. Then, the PG will ask the customer to confirm his/her payment request. Once it receives the confirmation, the PG will update the accounts of both the customer and merchant to complete the transaction.

In order to resolve potential conflicts in SIMPA, the OI and PI must be linked in order to allow the customer to demonstrate that the PI is solely intended for a particular order. The integrity of the OI and PI can be protected by the use of a dual signature (DS) scheme [114].

SIMPA provides a more secure platform for mobile payment services, and it is possible to implement strong security payment protocols on it. However, there is no way to verify it since the performance and security analyses are not provided in the paper.

In recent years, some schemas to perform payments in SIP have been proposed but they present few limitations at the moment when a person wants to pay for access to real-time services since either they are not suitable for micro-payments or they do not have sufficient security during payment information exchange. In order to overcome these limitations, the authors of [152] proposed a new Lightweight

Payment SIP (LP-SIP) protocol that can be used to pay for real-time services. The proposed protocol supports payment according to different models like pay-per-time, pay-per-data, session-based charging, and pay-as-you-watch.

LP-SIP is based on the use of e-coins that are obtained prior to accessing the SIP-based real-time service. These e-coins are charged to the client's account when he/she withdraws them. In other words, the proposed protocol is based on tokens and prepayment. The client acquires e-coins from a broker, and the e-coins can be used with any vendor registered with that broker. Hence, any entity has to complete a registration process with the broker to participate in the system.

Researchers have conducted a deep comparison between LP-SIP and SIPCoin since previous studies indicate that SIPCoin is the only proposal that can be considered to realize real-time micro-payments to access multimedia services based on SIP. Both proposals were analyzed regarding the number of messages exchanged between the different parties of the system (Client C, Vendor V, Broker or Payment provider B) during the payment phase and for the different cryptographic operations involved to make the payment. As a result, we can conclude that LP-SIP provides divisibility, better security properties (including efficiency), and reduced participation of the broker compared to SIPCoin, thereby making it suitable for real-time payment.

Symmetric cryptography provides the security for LP-SIP. However, in other processes where there is no previous contact between entities, such as the client and vendor registration processes, asymmetric encryption and digital signatures are used. The security analysis of LP-SIP shows that the proposed protocol satisfies, in terms of security, the following properties: integrity and authentication, confidentiality, and non-repudiation. Also, the proposed protocol is resistant to some attacks, such as replay attacks, bogus e-coins, double-spending/overspending, and double deposit/over deposit.

### *3.1.5 Communication Restriction*

In a restricted connectivity scenario, two entities of the model can not communicate with each other directly and must do so through a third party. The research published by [22] discusses various proposed scenarios used in mobile commerce, some of them with communication restrictions that are described as follows:

- *Kiosk Centric* (as shown in Fig. 3.32): The merchant behaves like a proxy to permit communication among the client and the issuer owing to the communication restriction that impedes direct communication among both entities. In this model, the client can only connect to the Merchant through a short-range link (such as Infrared, Wi-Fi, Bluetooth, NFC, or RFID).
- *Client Centric* (as shown in Fig. 3.33): Unlike the previous model, in this scenario, the merchant is not able to communicate directly with the acquirer. The client behaves like a proxy to permit communication between the entities. As in the previous scenario, the client communicates with the merchant via a short-range link.

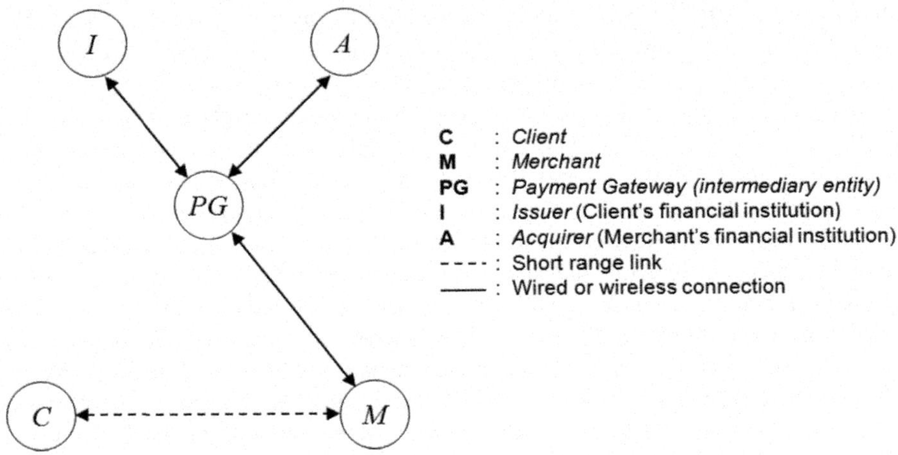


**Fig. 3.32** Scenario with merchant behaving like a proxy: the Kiosk centric model [78]

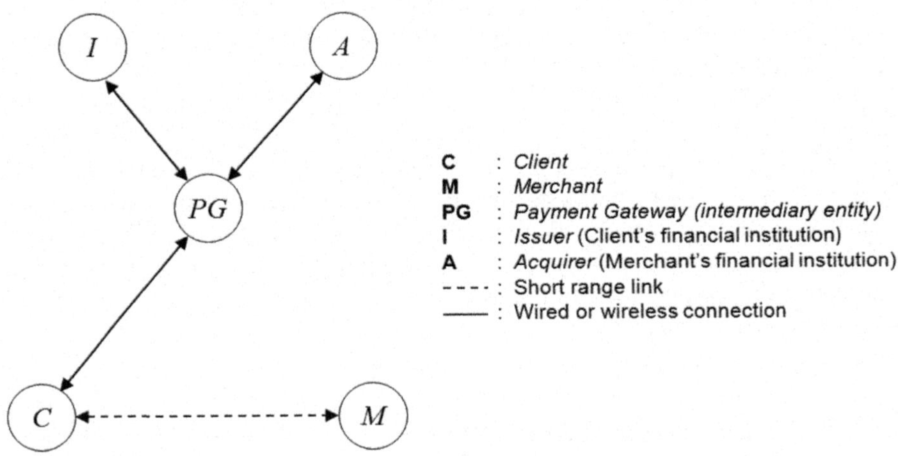


**Fig. 3.33** Scenario with client acting as proxy: client centric model [78]

- *PG Centric* (as shown in Fig. 3.34): The PG behaves as an intermediary between the client and the merchant owing to the lack of direct communication between the client and the merchant.

The aforementioned scenarios introduce new challenges for the designers of MPSs, who seek to provide the same security levels as those possible in full connectivity scenarios. Hence, security can help support this type of scenario and increase the ability to stimulate the creation of MPSs based on connectivity restricted scenarios that could be used in the real world. Some proposals based on these restricted connectivity scenarios can be found in the literature and will be presented here.

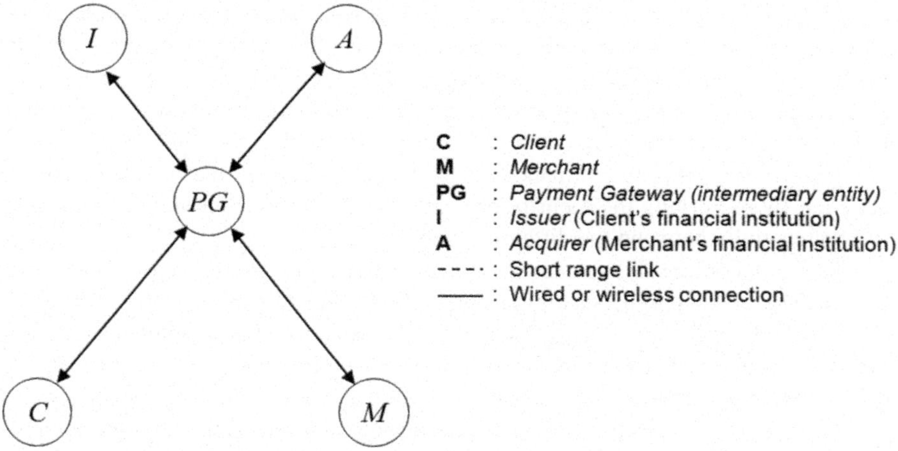


**Fig. 3.34** Scenario with PG as the intermediary between the client and the merchant [78]

In 2006, [72] proposed a secure protocol for an MPS based on a kiosk centric case mobile scenario. The proposed protocol (hereafter referred to as the KCMS protocol) utilizes symmetric key operations, which need low computational power and are suitable for MPSs where the customer cannot communicate with the issuer due to the absence of Internet access on his/her mobile device and the cost of implementing other mechanisms of communication between them are high.

The proposed protocol consists of four sub-protocols: *registration* (which involves the client, the merchant, and the issuer), *purchase* (which is carried out among the client and the merchant over the wireless channel), *withdrawal* (which occurs between the merchant and the PG through a secure wired channel), and *deposit* (which occurs between the PG and the acquirer trough a secure wired channel). Despite the separation of the protocol into several sub-protocols, which does not enhance overall performance, there is a reasonable abort-handling mechanism in comparison with the integrated one, the authors claim that the performance is adequate when compared with other protocols. Also, the proposed protocol satisfies the transaction security properties that other MPSs that utilize public key cryptosystems provide.

Later in the same year, [174] proposed an anonymous protocol for an MPS based on a KCMS. It protects the real identity of the clients during the purchase and uses a digital signature scheme with message recovery (discussed in detail in Chap. 4) that uses self-certified public keys that reduce communication costs when compared to the certificate-based signature schemes.

Using the Client Centric Model, the authors of [70, 71] presented two proposals. The first one is an anonymous payment protocol for a Client Centric Mobile Scenario where the merchant has no direct communication with the acquirer due to the absence of Internet access in his/her infrastructure and the lack of affordability of other communication technologies due to their associated inconveniences and costs [70]. The symmetric key operations employed in the proposed protocol for all engaging parties reduce both the setup cost for the payment infrastructure and the transaction cost. Moreover, it supports credit card and debit card transactions and protects the real identity of the clients during the purchase. As a result, the proposed payment protocol represents an alternative to other MPSs, which have restrictions regarding a mandatory connection between the merchant and the acquirer, by preserving the security properties as if it is working in a scenario with full connectivity between the different entities.

The second proposal is a novel anonymous protocol for an MPS based on a Client Centric Model that employs a digital signature scheme with message recovery using self-certified public keys that allow a client to make purchases without disclosing private information [71]. The proposed payment protocol supports credit card and debit card transactions and protects the real identity of a client during the payment. As a result, the proposal illustrates how a merchant can sell goods in a secure way even if the merchant cannot directly communicate with the acquirer due to the preservation of the security properties.

Taking into consideration the advances in VANETs, [73] proposed a secure protocol for online payments in a vehicle-to-roadside restricted scenario in VANETs where the client cannot directly communicate with the issuer. The proposed protocol, known as the Kiosk Centric Model Payment protocol for VANETs (from now on referred to as the KCM-VAN protocol), employs the authentication encryption scheme [197] proposed. Moreover, it supports both credit card and debit card transactions, protects the real identity of a client during the payment, and can be used with an AU (including portable devices). The five entities that make up the KCM-VAN and their interactions are shown in Fig. 3.35.

The security analysis the authors of the KCM-VAN protocol presented demonstrates that it can withstand replay and impersonating attacks. Also, the authors argue that the proposal is efficient and can be deployed for the practical scenarios of restricted connectivity in VANETs.

An improved version of the KCMS protocol [77] but applied to the same restricted scenario in VANETs was proposed by [77], called the KCMS-VAN protocol, which uses symmetric key operations for all engaging parties to decrease both the setup cost for the payment infrastructure and the transaction cost. Besides, it could be used with a portable device attached to an AU, preserves the real identity of the client throughout a payment, and permits both credit card and debit card transactions.

In order to perform purchase transactions, the KCMS-VAN protocol needs a software (from now on referred to as the KCMS-VAN wallet) on the client side. Figure 3.36 shows snapshots of the main screen and main menu of the KCMS-VAN wallet on both Nokia™ N95 (top) and Palm™ T|X devices (bottom).

The empirical results from the performance evaluation of the implementation presented in the work have proven that the proposed KCMS-VAN protocol can conduct payment transactions over a wireless network. The implementation demonstrates that a KCMS-VAN payment transaction can be completed within an average of 6.84 s using a Nokia N95 device and 5.66 s using a Palm T|X device.

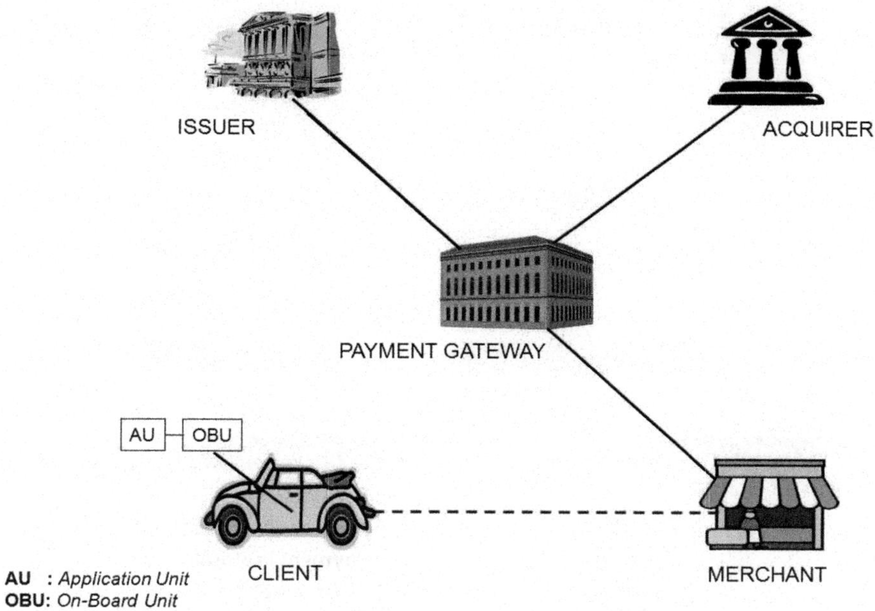


**Fig. 3.35** Operational model for KCM-VAN payment protocol [73]

Based on the work [73] proposed and the open problems discussed in [77], [104] proposed an efficient and secure mobile payment protocol for restricted connectivity scenarios in VANETs. The proposed payment protocol was designed for data volume transactions in the restricted connectivity scenario of VANETs where the vehicle cannot directly communicate with the bank and employs the self-certified key agreement to generate the symmetric encryption keys, which combine the advantages of both public key cryptography and symmetric key cryptography. Thus, the session key is fresh for each payment transaction, thereby providing more robust security protection. Also, self-certified public keys (SCPKs) on an elliptic curve are used in order to avoid the requirement of a large public key infrastructure (PKI) and to achieve efficient performance in contrast to other public key cryptosystems. Moreover, the proposed protocol supports not only data volume transfer-based applications but also one-time event applications.

The performance and security analysis demonstrate that the self-certified key agreement employed to establish the shared key between two participants (without additional message exchanges) results in efficient symmetric encryption. Therefore, the communication and computation costs are reduced. Also, properties such as fair exchange, user anonymity, and payment security are satisfied in the proposed protocol.

As an extended version of the KCMS payment protocol [72] proposed, [159] proposed a Secure Kiosk Centric Mobile Payment protocol using symmetric key techniques. The main difference between both versions of the KCMS payment

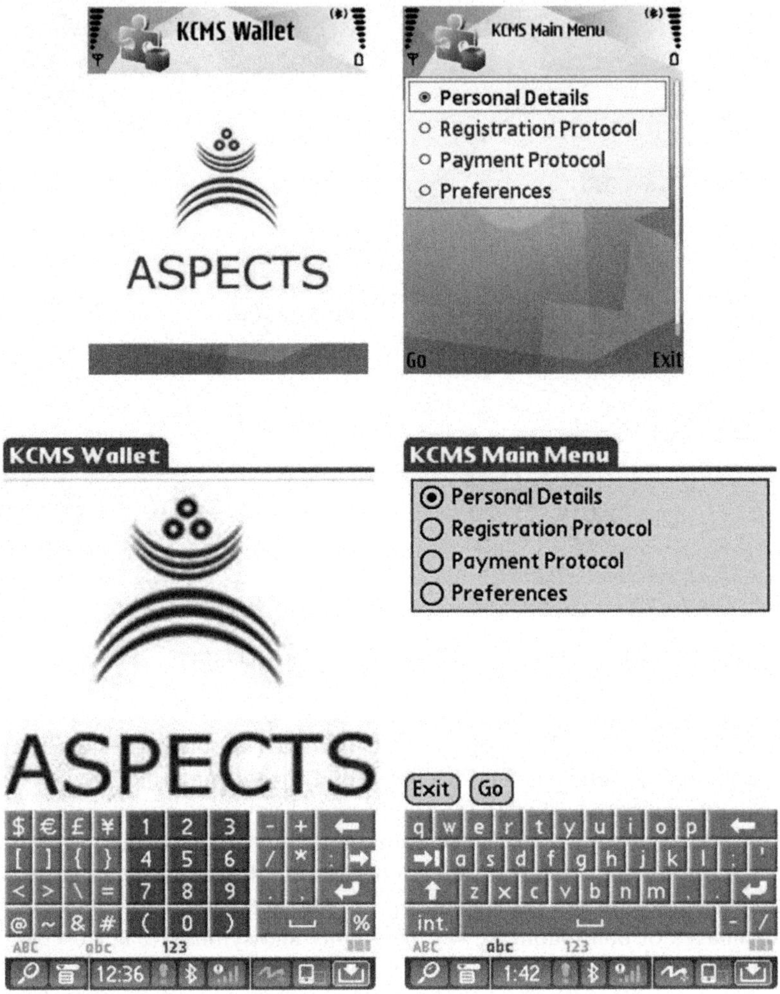


**Fig. 3.36** Snapshots of main screen and main menu of the KCMS-VAN wallet on both Nokia N95 and Palm T|X devices [77]

protocol is that the original version satisfies only one non-repudiation service (non-repudiation of origin (NRO)) while the newest version satisfies the other existing non-repudiation services [54]: (a) non-repudiation of receipt (NRR), (b) non-repudiation of submission (NRS), and (c) non-repudiation of delivery (NRD).

In the context of the VANET, [78] designed and implemented a lightweight secure payment protocol (employing symmetric key operations) for those scenarios in VANETs and other mobile environments where the merchant is not able communicate directly with the acquirer to process the payment request. The proposed protocol, known as the Client Centric Model Payment protocol for VANETs

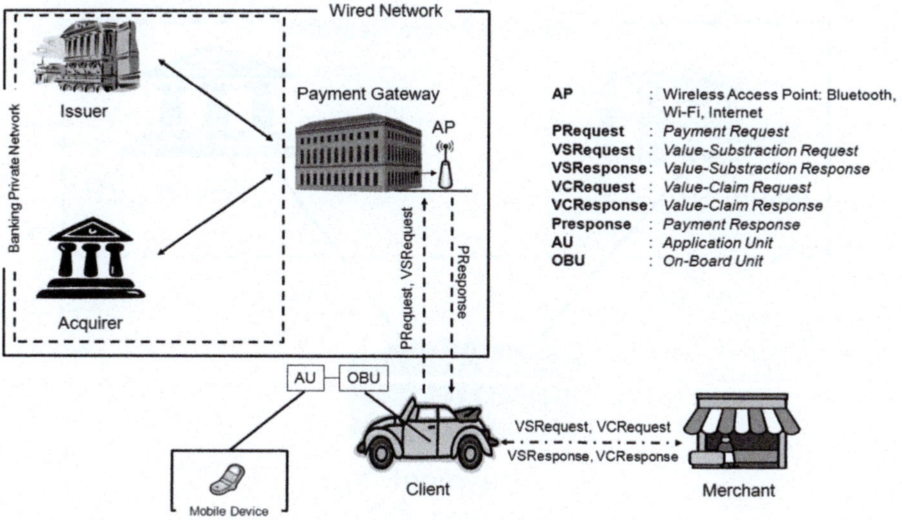


**Fig. 3.37** Proposed CCMS-VAN design architecture [78]

(henceforth referred to as the CCMS-VAN protocol), can be used by a portable device attached to an AU (including portable device), supports both credit card and debit card transactions, and protects the real identity of the client during the payment. The architecture for the proposed payment CCMS-VAN protocol and the interactions of five entities that compose it are depicted in Fig. 3.37.

The security analysis the authors of CCMS-VAN protocol presented shows that it can withstand replay and key guessing attacks. The implementation of the CCMS-VAN payment protocol demonstrated that a payment transaction can be completed within average of 8.34 s using a Nokia N95 device.

Recently, [74] proposed the design of an anonymous secure payment protocol based on the PG Centric scenario that eliminates the restriction of those MPSs on the Full Connectivity Scenario that require direct communication between a client and a merchant for authentication purposes. Moreover, the proposed payment protocol supports both credit card and debit card transactions, protects the real identity of the client during the purchase, and employs symmetric key operations, which require low computational power and can be processed faster than asymmetric operations. The five entities that make up the PCMS and their interactions are shown in Fig. 3.38.

The performance analysis (in terms of the number of cryptographic operations of the involved parties and computation costs) of the proposed secure PCMS demonstrates the proposed protocol's superior performance over the KSL (proposed by [93]) and LMPP (proposed by [44]) payment protocols in terms of efficiency even when using a scenario with communication restrictions [74].

The PCMS protects the real identity of the client during the purchase by using a nickname (a temporary identity known only to the client and the issuer) instead of

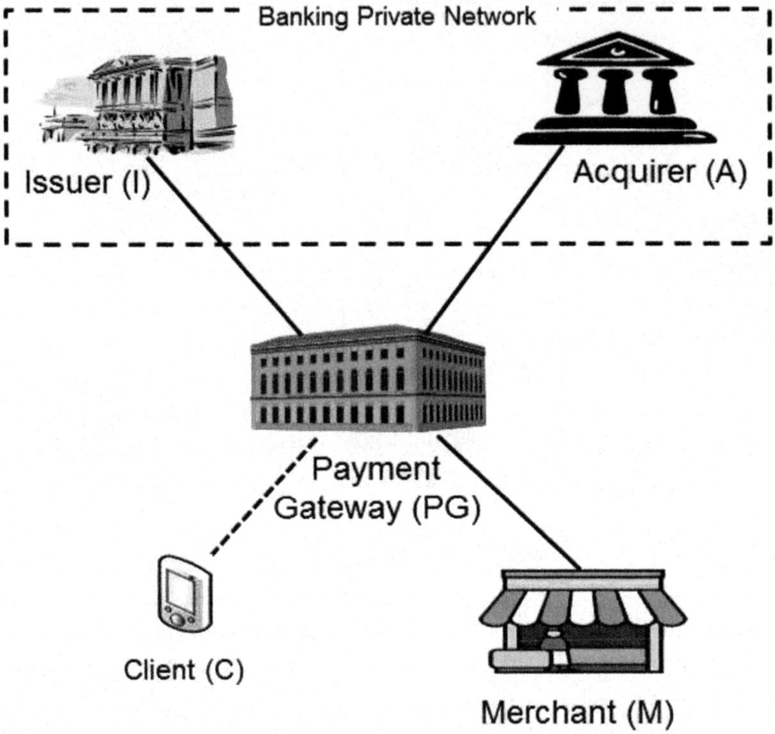


**Fig. 3.38** Operational model for PCMS payment protocol [74]

the client's real identity. The symmetric cryptography and message authentication code used in the proposed protocol ensure the confidentiality and integrity of the messages transmitted in each transaction. Also, the PCMS uses a timestamp included in the transmitted message to prevent replay attacks and to guard against key guessing attacks.

The authors of [75] presented the implementation of the PCMS payment protocol. The results demonstrate that the implemented protocol (which deploys lightweight and secure cryptographic algorithms) can complete a payment transaction within an average of 11.68 s using a Nokia N95 device.

### *3.1.6 Mobile Agent Technology*

A mobile agent is a kind of special agent that possesses the basic attributes of an agent and the important feature of mobility. It can be defined as an autonomous program that can travel smoothly from computer to computer in a network at times and to places of its own choosing. Moreover, the mobile agent can clone or generate a child agent that has the same nature as the parent agent if needed. The state of the running program is saved when it is transmitted to the destination. At the destination, the program resumes and continues processing from the saved state [103, 154].

A mobile agent can shift autonomously and automatically both in a heterogeneous network and in a distributed computer environment. Moreover, it can fulfill other tasks, such as carrying information or looking for appropriate information resources, processing information immediately, helping users with information delivery, web inquiry, data and knowledge discovery, information conversion, and so on [103]. Moreover, mobile agents have several advantages in the development of various services in smart environments in addition to distributed applications such as the following [154]:

- **Reduced Communication Costs:** Since distributed computing requires interaction between different computers through a network, the latency and network traffic of interactions often seriously affects the quality and coordination of two programs running on different computers. Figure 3.39 illustrates how mobile agent technology enables remote communications to operate as local communications. That is, if one of the programs is a mobile agent, it can migrate to the computer on which the other is running to communicate with it locally.
- **Asynchronous Execution:** Once a mobile agent has migrated to the destination-side computer, it does not need to interact with its source-side computer. Therefore, the agent can continue processing at the destination even if the source is shut down or the network between the destination and source is disconnected. This is useful in unstable communication, including wireless communication, scenarios in smart environments.
- **Direct Manipulation:** A mobile agent is executed locally on the computer it is visiting. Therefore, it can directly access and control the equipment of the computer as long as the computer allows it to do so. This is helpful in network

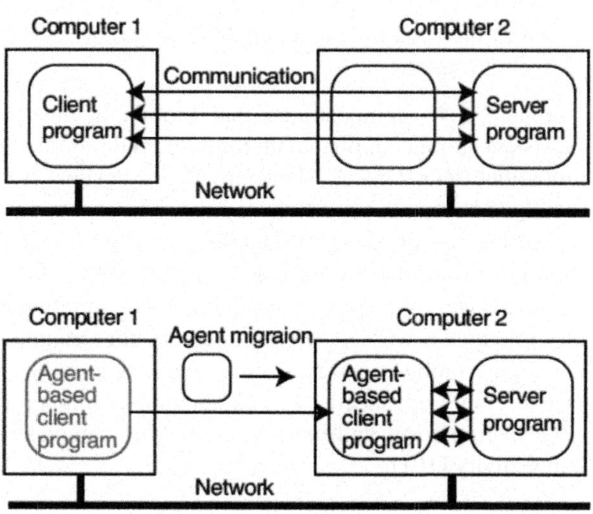


**Fig. 3.39** Reduced communication with mobile agent technology [154]

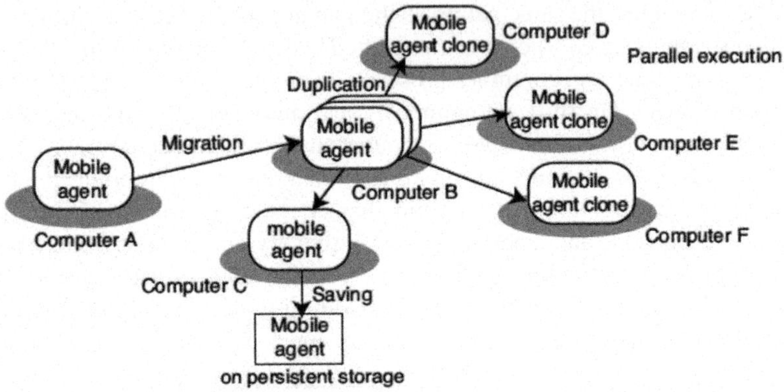


**Fig. 3.40** Functions of mobile agents in distributed system [154]

management, particularly in detecting and resolving device failures. Installing a mobile agent close to a real-time system may prevent delays that network congestion can cause.

- **Dynamic-Deployment of Software:** As a mechanism for the deployment of software, mobile agents are useful because they can decide their destinations and their code and data can be deployed dynamically at that destination when needed. This is useful in smart environments because they consist of computers whose computational resources are limited.
- **Easy-Development of Distributed Applications:** Most distributed applications are made up of at least two programs (i.e., a client-side program and a server-side program) and often spare codes for communication, including exceptional handling. However, because a mobile agent can carry its information to another computer, it is possible to write a single program to define distributed computing. Moreover, since a mobile agent program does not have to define communication with other computers, standalone programs can be modified easily as mobile agent programs.

Figure 3.40 demonstrates that mobile agents can save themselves through persistent storage, duplicate themselves, and migrate themselves to other computers under their own control. Therefore, they can support various types of processing in distributed systems.

Mobile agents are self-contained entities capable of autonomously roaming the Internet and launching user-assigned tasks. This paradigm brings forward the creative idea of moving user-defined computations towards network resources and provides a wholly new architecture for designing m-commerce applications. Deploying mobile agents in m-commerce can reduce unnecessary network traffic, provide more advanced services, support automation of decision making, reduce participation costs, and improve trading efficiencies while tolerating poor network connectivity [101].

An agent-based payment system refers to those current payment systems that improve its ability to execute transactions in a wireless environment by applying mobile agent technology. In this kind of payment system, a mobile client sends an agent, which includes information related to the payment needed to execute a transaction in a merchant's environment, over a fixed network and reduces connection cost for the clients [92].

Guided by the SET rules and using the mobile agent paradigm, [150] proposed an agent-based SET payment system (called SET/A) that resolves the problem of implementing the SET protocol [114] in wireless environments. The proposed solution uses a mobile agent that executes a transaction on behalf of a client outside the client's environment.

In SET/A, the client needs to be connected to the Internet for only two short periods of time: (a) to send the agent containing the client's request to the merchant at the beginning of the transaction, and (b) to receive the agent containing the result of the client's request at the end of the transaction. However, performing the SET operations on the merchant's side makes the transaction vulnerable to attack because the merchant's server is considered to be a hostile environment [187]. To address this security problem, the SET/A protocol suggests running the agent in a tamper-proof environment or a secure coprocessor to protect the agent against malicious merchants. This means that the agent would migrate to a protected (hardware) environment securely attached to the merchant's server, and all the confidential data would be handled inside this environment. That being said, this solution would increase the security level but at the cost of an additional investment in hardware from the merchant [191].

To resolve the issues of SET/A identified in this section, the authors of [187] proposed another agent-based payment system for mobile computing on the Internet, namely SET/A+, which removes the security limitation of the agent's execution environment on the merchant's server by adding a trust verification center in the payment system for mobile computing. SET/A+ was designed to be compatible with SET. It only requires significant modifications on the cardholder's side. The merchant software could remain unchanged since its interaction with the agent is largely the same as it would be with the cardholder. The only exception is that the software must be aware that now it is communicating with an entity residing on a host other than the cardholder's [191].

SET/A+ operates in a larger scenario than that of SET/A and includes brokering and negotiation phases, which require the capability of a mobile agent in the SET protocol. SET/A+ is suitable both for a payment system in which a single merchant is involved, and for a payment system with multiple merchants [92, 191].

Recently, [105] proposed a new secure lightweight mobile bill payment protocol with the assistance of an intermediary acting as an agent to collect and process clients' payment requests. The proposed protocol aids clients and merchants by providing the assistance of more reliable intermediaries that may reside in a fixed network. Moreover, the proposed protocol deploys lightweight cryptographic operations (e.g., symmetric key and MAC operations) that make it more lightweight than existing protocols and suitable for making bill payments in wireless environments.

To deal with the unreliability of wireless networks, the proposed protocol has intermediaries (located on reliable, fixed networks) that act as either a client agent or merchant agent. Therefore, after making a request, a client or a merchant may disconnect from the network due to a wireless network problem since the intermediary can process requests further and forward them to related parties. It is important to note that, when the response is sent to the intermediary, the intermediary can keep the response and forward it to the client when he/she reconnects to the network [105].

### 3.1.7 *Wireless Application Protocol (WAP)*

In 2008, [117] proposed a secure mobile payment model based on the WAP. The model adopts bridge CA architecture instead of hierarchical CA architecture to overcome the weaknesses of collaboration between the participants in the m-commerce transactions and demonstrates improved interoperability, scalability, and robustness. Since each participant organization varies from the others in m-commerce, it is not reasonable to expect that each participant will subordinate its trust model to a third party, thereby making the hierarchical model unsuitable for mobile payments.

The transaction process in the mobile payment model based on the WAP, which uses the previously described bridge CA authentication model, is illustrated in Fig. 3.42. It is important to note that, once the cardholder has selected the items to be purchased through the WAP browser, he/she obtains the Transaction Identity number (TID) from the merchant and then the transaction begins as follows [117]:

**Fig. 3.41** Bridge CA authentication model [117]

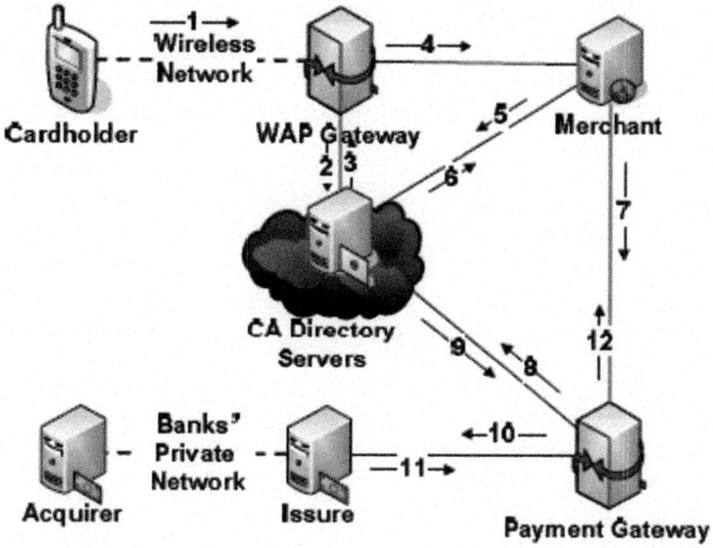


**Fig. 3.42** Process of transaction in the proposed mobile payment model based on WAP [117]

I. **Step 1:** The cardholder sends to the WAP Gateway (WG) the Purchase Information (PUI), encrypted by the WAP gateway's certificate. The PUI is composed by the Order Information (OI), encrypted Payment Instruction (PI), two signatures, and the URL of the user's certificate, encrypted by the WAP gateway's certificate.
II. **Steps 2–4:** The WG forwards the received PUI to the merchant.
III. **Steps 5–6:** Once the PUI is received by the merchant, it is decrypted from the WG. Then, the message is verified using the cardholder's certificate.
IV. **Step 7:** A petition is sent to the PG, including a TID and a digest of the OI and user's encrypted PI.
V. **Steps 8–10:** PG obtains the cardholder's PI and merchant's TID. Afterward the verification of the identification and message, PG sends an authorization request to the issuer through the bank's private network. The transfer of capital is processed by the bank via the finance intranet.
VI. **Steps 11–12:** Afterward the transfer between the banks, the bank sends the catch token to the PG. The PG signs the token digitally and sends both the token and signature to the merchant.

The authors' discussion in the work clearly shows that the bridge CA authentication model can be configured to work well both over wired and wireless networks.

# Chapter 4
# Security in Mobile Payment Systems

Security is the biggest issue in the field of m-commerce because no one will trust m-commerce without secure commercial information exchange and safe electronic financial transactions over mobile networks. Therefore, various mobile security procedures and payment methods have been proposed and applied to m-commerce [85].

Cybercriminals have targeted unsuspecting mobile payment users. This problem is especially critical in emerging markets where cybercrime-related legal frameworks and enforcement mechanisms are not well developed. There are many instances of phishing attacks targeting mobile money users [90].

Since MPSs include the transmission of sensitive data, it is imperative that users and MPS designers be very careful with security issues. Thus, to encourage widespread use and customer acceptance of mobile payments services, both perceived and technical levels of security should be high. For customers, privacy should not be compromised, and there should be no possibility of financial loss. It is important to keep in mind that security always requires an overall approach. Ultimately, a system is only as secure as its weakest component, and securing networking transmission is only one part of the equation [98].

## 4.1 Security Requirements

The concrete security requirements of MPSs vary depending both on their features and the trust assumptions placed on their operations. In general, however, MPSs must satisfy the following transaction security properties [9, 85, 183]:

- **Authentication** is concerned with verifying the identities of parties in a communication and ensuring that they are who they claim to be. Therefore, in the system, every engaging party should be able to authenticate the party with whom he/she is trying to communicate.



J. Téllez, S. Zeadally, *Mobile Payment Systems*, Computer Communications and Networks, DOI 10.1007/978-3-319-23033-7_4

- **Confidentiality** involves ensuring that a message is not revealed to any unauthorized parties and only can be read by the sender and intended recipient of the message.
- **Integrity** is concerned with ensuring that the content of the messages and transactions cannot be altered, whether accidentally or maliciously, during the transmission.
- **Authorization** involves the procedures that must be in place to verify that the user can make the requested purchase.
- **Non-repudiation** is concerned with providing mechanisms to guarantee that a party cannot deny that he/she has participated in a transaction. Non-repudiation differs from authentication in that the latter only needs to convince the other party involved in a communication of the validity of an event while the former must prove to a third party the truth of the event. The primary purpose of non-repudiation is to protect a user's communication against threats from other legitimate users rather than from unknown attackers. Depending on the scenario of an electronic transaction, different non-repudiation services can be employed. The following non-repudiation services are required in order to establish the accountability of the actions of the originator and the recipient [54]:

  - *Non-repudiation of Origin (NRO):* This intended to protect against the originator's false denial of having originated the message.
  - *Non-repudiation of Receipt (NRR):* This intended to protect against the recipient's false denial of having received the message.
  - *Non-repudiation of Submission (NRS):* This intended to provide evidence that the originator submitted the message for delivery.
  - *Non-repudiation of Delivery (NRD):* This intended to provide evidence that the message has been delivered to the recipient.

The transaction security properties described above define what the authors of [128] call *end-to-end security property*, which ensures that all information sent from the user to the entity with which he/she is communicating is secure at all times during the payment transaction made using mobile devices.

An additional property to the central properties described in this section and which is required by any secure payment system is called *"accountability"*, which refers to ability to demonstrate that the parties involved in the system are responsible for their transactions. Accountability property is acknowledged as a high-level security property that satisfy the previously mentioned central transaction security properties. To fulfill the accountability property, a protocol must satisfy all of the transaction security properties stated earlier [92].

Cryptography plays a central role in the satisfaction of the transaction security properties already mentioned. Also, cryptographic techniques are essential tools in securing payment protocols over open, insecure networks. Therefore, a cryptographic implementation is necessary for the design of a payment security architecture that can be implemented practically on resource-limited mobile devices and is compatible on wireless networks. The employment of a cryptographic strategy contributes to the security a monetary architecture requires, but it often imposes computational requirements that cannot be met in resource-limited environments [200]. Section 4.2 provides a brief review of the most commonly used cryptography schemes for secure communications among engaging parties.

## 4.2 Basic Concepts in Cryptography

Cryptography is the art and science of securing messages so that unintended audiences cannot read, understand, or alter them. Its main purpose is to take ordinary plain text messages and scramble them or convert them to cipher text using various algorithms. The recipient can, in turn, decipher the scrambled message and recover the original plain text message using a matching decryption algorithm. For added security, both encryption and decryption algorithms rely on one or more keys. Keys are generally kept secret (or at least one key is in the case of public key cryptography) and are combined with the messages that need to be encrypted or decrypted. Thus, no one can read the content unless they have access to the secret key even if the encryption and decryption algorithms are made public. Moreover, since the encryption algorithms rely on mathematical properties, it is difficult to try to guess the key or the content of the encrypted message [153].

### *4.2.1 Secure Sockets Layer (SSL)*

SSL, originally developed by Netscape, is defined as a cryptographic protocol designed to provide security across a communication network (like the Internet) that involves a client and server. Typically, SSL only authenticates the server to the client (unilateral authentication), and the client can communicate with the server anonymously. The security of the communication channel is provided using various cryptographic methods, such as cryptographic algorithms, message digest functions, and digital signatures.

Many websites use SSL for secure areas of their sites, such as user account pages and online checkout. Usually, when a user is asked to "log in" on a website, the resulting page is secured by SSL. Most browsers (Internet Explorer, Mozilla Firefox, Safari, Google Chrome, etc.) support SSL, and many websites use the protocol to obtain confidential user information, such as credit card numbers. By convention, URLs that require an SSL connection start with *https:* instead of *http:*.

### *4.2.2 Symmetric Cryptography*

Secret key cryptography, also known as symmetric cryptography (which includes symmetric encryption and message authentication codes), relies on the use of the same cryptographic keys for both the encryption of plain text and the decryption

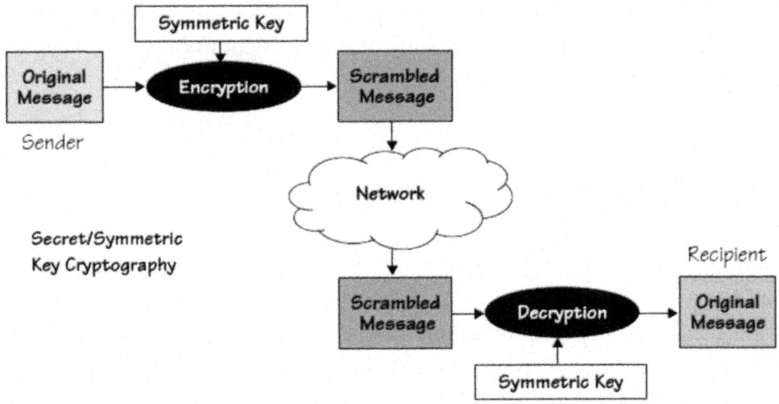


**Fig. 4.1** Private/symmetric key encryption [153]

of cipher text. A secret is shared between two parties (known as the sender and receiver) that want to communicate safely without revealing the details of the message (see Fig. 4.1). Message confidentiality, message integrity, and party authentication are provided by this technique [72].

The following example describes a two-party communication with the symmetric key encryption: Alice (a sender) wants to send a message to Bob (a receiver). She does not want any unauthorized person to read the message, so she encrypts it using her secret key, $K_i$, and sends it over a public or unsecure channel. For Bob to read the message, he has to have the same key, $K_i$, that Alice used to encrypt the message. An adversary can intercept the message but cannot read it. The message cannot be decrypted easily without the secret key. Using a secure channel makes the data more resistant to intercepting and tampering [4].

Symmetric cryptography works well if people can agree on a secret key ahead of time. However, when a person tries to communicate over the Internet with people he/she has never met before, he/she might be faced with the following problem: how can he/she agree on a key with them if he/she does not already have a secure means of communication? Storing different keys ahead of time for every person he/she might one day want to communicate with does not sound very practical, and using the same key with more than one person raises non-repudiation issues. If several people use the same key to communicate with that person, how can he/she tell which one actually sent a message? Therefore, public key algorithms provide a much more straightforward way of dealing with key exchange, a particularly important issue over the Internet where people constantly interact with new people. However, despite the complexity of key management, symmetric key operations are more suitable for wireless networks than asymmetric ones due to their required processing time and lower computational requirements. For this reason, while public key algorithms are often used to exchange secret keys at the beginning of secure sessions, data transmission during the session actually relies on symmetric cryptography using the exchanged key [72, 153].

Some examples of symmetric encryption systems are the following: International Data Encryption Algorithm (IDEA) [96], Advanced Encryption Standard (AES) [131], Data Encryption Standard (DES) [130], and Triple Data Encryption Standard Algorithm (Triple DES) [130].

#### 4.2.2.1 Message Authentication Code (MAC)

One of the goals most strongly associated with cryptography is privacy, which should protect against passive attacks (eavesdropping). However, protection against active attacks (falsification of data and transactions), which is known as message authentication, is arguably even more important. Message authentication (also known as data-origin authentication) is a procedure that allows communicating parties to verify that the received message is authentic.

MACs use symmetric key cryptography to protect against both accidental and malicious message tampering. The process begins with the exchange of a symmetric key at the beginning of a session, typically using public key cryptography. Afterward, the exchange is applied to message digests, and both the result and the message are sent to the receiver, who applies the same algorithm and symmetric key to the received message to verify both its authenticity and integrity. If the comparison between the sender's MAC and the receiver's MAC indicates that they match, the message has not been altered and the source is authentic. Otherwise, the message has not arrived in the same condition in which the sender sent it. Figure 4.2 shows the MAC procedure.

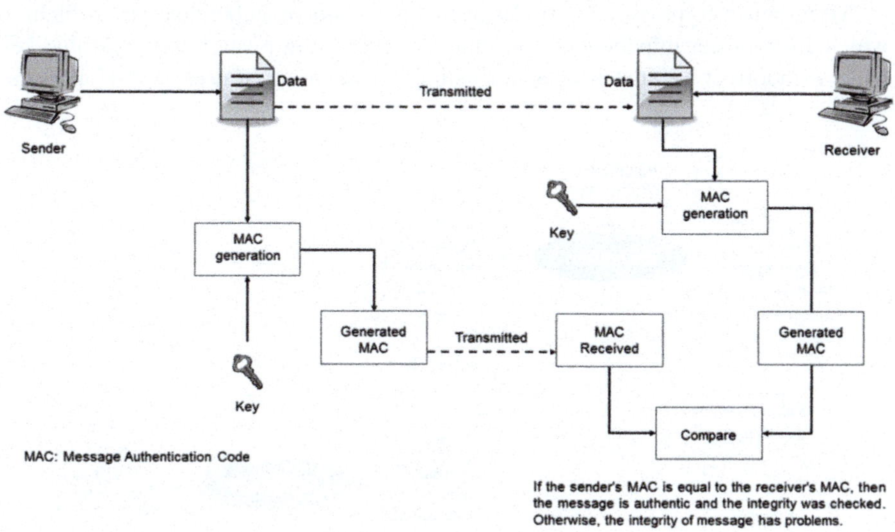


**Fig. 4.2** Message Authentication Code (MAC) procedure

### 4.2.3 Public Key Cryptography

Public key cryptography or asymmetric cryptosystems, which Diffie-Hellman introduced in 1976, is a class of cryptography that permits users to communicate safely without having prior access to a shared secret key. In this technique, each party receives a pair of keys: one of them is referred to as the public key and the other as the private key. Each public key is published while the private key remains secret. The interesting property of this pair of keys is that messages encrypted with one key require the other key for decryption [72, 153].

Unlike symmetric cryptography, is not necessary to share the secret key between the sender and receiver in public key cryptography. All communications involve only public keys, and no private key is ever transmitted. Also, users no longer need to trust communication channels to be secure against eavesdropping or betrayal. The only requirement is that public keys are associated with their users in a trusted manner (for instance, in a public directory maintained by a System Authority (SA)) [200].

In public key cryptography the two-party communication encryption works as follows (see Fig. 4.3): Alice (a sender) wants to send a message to Bob (a receiver). She does not want any unauthorized person to read it, so she obtains Bob's public key from the public directory and encrypts her message using that key before sending it to Bob over a public or unsecure channel. An adversary can intercept the message but not read it since he/she does not have Bob's private key, which is necessary to decrypt Alice's message. Once Bob has received Alice's message, he decrypts it using his private key.

Asymmetric cryptosystems suffer from a well-known authentication problem in which an imposter impersonates an innocent user using a valid cryptographic key but it is incorrect because it does not belong to the innocent user [72]. Hence, use

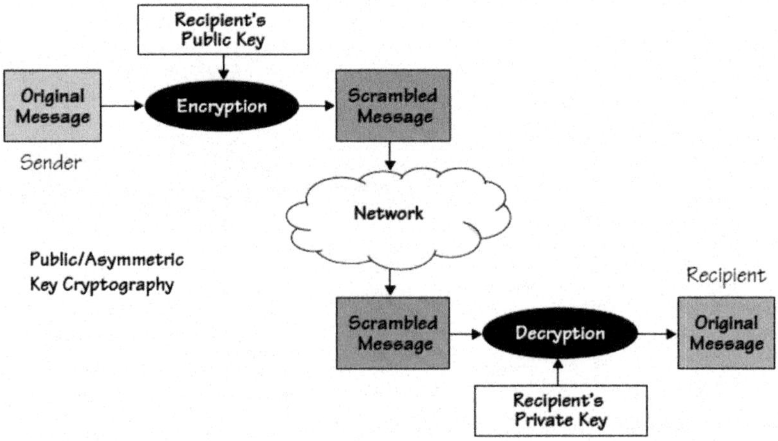


**Fig. 4.3** Public/asymmetric key encryption [153]

of certificates with every public key is necessary to overcome this issue. However, since users have to connect to the SA in order to verify the certificate, a high burden is placed on them [174]. The following section provides an introduction to digital signatures.

RSA (which stands for the first letter in each of its inventors' last names: Ronald Rivest, Adi Shamir, and Leonard Adleman) is the most popular algorithm for public key cryptography [148]. It works with variable key sizes: longer keys require more memory and processing but provide stronger security. Many security protocols are based on the RSA algorithm.

In modern communication, mobile phone subscribers are highly concerned with personal privacy and security. Moreover, modern mobile phones are increasingly being used for more services that require modern security mechanisms such as the RSA public key cryptosystem. However, given the limit of mobile devices' processing capabilities, mainstream mobile manufacturers are still unable to apply advanced security protocols to mobile devices. To implement the RSA algorithm and apply many advanced security protocols to mobile networks, some variants of the RSA cryptosystem (such as CRT, Multi-Prime RSA, Multi-Power RSA, Rebalanced RSA, R-Prime RSA, and Hwangs method) can be used efficiently [59, 69].

Recently, the use of Elliptic Curve Cryptography (ECC) algorithms into mobile devices is gaining acceptance. ECC algorithms rely on different mathematical properties that allow for shorter keys, a particularly appealing feature on mobile devices. However, widespread acceptance of new cryptographic algorithms takes time. Section 4.2.4 briefly discusses ECC algorithms.

#### 4.2.3.1 Digital Signatures

The digital signature (which is analogous to a handwritten signature on a paper document) is a mathematical scheme to protect data integrity and to achieve authentication and non-repudiation. A digital signature uses a private key to encrypt messages and a public key to decrypt messages. By comparing the signature to the original message, it is possible to determine if the message has been changed and if it has been signed using a particular key pair. Three main properties are required in digital signature schemes: verification of the sender at the time of the signature, authentication of the content at the time of the signature, and verification by third parties to resolve disputes between the sender and receiver.

The digital signature concept works as follows: Alice (a sender) is supposed to send a message to Bob (a receiver). She must ensure that the message is unchanged from what she sent and that she is the sender. To achieve these goals, Alice uses her private key from a key pair to encrypt the message and produce a digital signature. Afterward, Alice sends the message and digital signature to Bob, who then decrypts the received message using Alice's public key to verify the digital signature. If Bob is able to decrypt the message, he knows the message was encrypted with Alice's private key. Bob makes a message digest (the hash value $A$) from the received

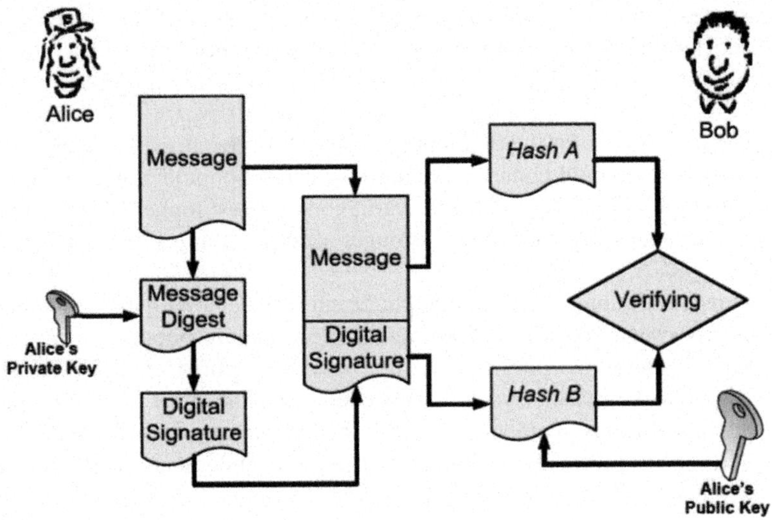


**Fig. 4.4** Digital signature [200]

message to compare it with the hash value *B* decrypted from the digital signature with Alice's public key. If the comparison of both hash values is successful, the received message is successfully verified. Therefore, the message sent by Alice is unchanged and is exactly the message Bob received. Figure 4.4 illustrates the digital signature signing and verification process.

A digital signature scheme consists of the following three components [6, 200]:

- *A key generation algorithm* that produces pairs of keys, made up of a private signing key that should be kept secret and a public verification key that should be distributed to everyone who may wish to verify a digital signature.
- *The signing algorithm* that takes as input the private signing key and the message that needs to be signed and outputs a digital signature for that message.
- *The verification algorithm* that takes as input the public verification key, a digital signature, and, possibly, the message that was signed. It either produces a valid output, which indicates that the given digital signature is the correct signature for that message, or an invalid output, which indicates that the given digital signature is not the correct signature for that message.

#### 4.2.3.2 Certificate Authorities

A Certificate Authority (CA) or a trusted third party (public or private) is an entity that issues digital certificates that certify a named subject's ownership of a public key. Thus, other parties can rest assured that the private key signatures correspond to the certified public key.

**Fig. 4.5** A digital certificate is signed by a CA and can be verified using the CA's public key. Typically, the certificate includes several keys; for example, one for digital signatures and one for key exchange [153]

The following example illustrates the importance of using certificate authorities (CAs). If an attacker *E* gains access to the directory where a recipient keeps Bob's public key and substitutes it with his, *E* could possibly intercept messages intended for Bob, alter them, and re-encrypt the resulting messages with Bob's actual public key. Thus, the recipient would not see the difference [153].

A CA can issue a public key certificate for Bob's prospective recipient signed with its own private key. The public key certificate would contain his/her name, public key, an expiration date, and some additional information (see Fig. 4.5). Afterwards, anyone can use the CA's public key (which is available to everyone) to verify the validity of Bob's public key certificate. If the check is successful, anyone should feel secure using Bob's public key to encrypt a message intended for him. The ISO standard format for certificates is known as X.509.

### *4.2.4 Elliptic Curve Cryptography*

ECC is an approach to public key cryptography based on the algebraic structure of elliptic curves over finite fields. Its security comes from the elliptic curve logarithm, which is the discrete logarithm problem (DLP) in a group defined by points on an elliptic curve over a finite field. This results in a dramatic decrease in the key size needed to achieve the same level of security offered in conventional public key cryptography schemes [26, 200].

Compared to RSA, ECC can provide the same level of security using a much smaller key size, thereby reducing processing overhead. This means the ECC offers faster computations, lower power and memory consumption, and network bandwidth savings. These properties are useful for mobile devices, which have limited memory and computing power and cannot expend much bandwidth for communication. Thus, ECC is theoretically more suitable for securing mobile banking than RSA [135, 200].

### 4.2.5 *Self-Certified Public Keys*

A digital signature provides authentication, data integrity, and non-reputation cryptographic services [100]. Traditional digital signature schemes are based on asymmetric techniques where each user has a public key (which anyone can know) and a private key (known only to the user). The user uses his/her private key to generate a signature for a given message, and the verifying party (the receiver of the message) uses the signer's public key to verify the signature. To ensure the authenticity of public keys and avoid the well-known problem of authentication that was mentioned in Sect. 4.2.3, keys must be certified. This can be achieved through two concepts: explicitly and implicitly verifiable certificates [139]:

- *Explicit Verifiable Certificate:* In this case, the authenticity of the public key requires a guarantee called a certificate (X.509 being one of the most used [80]) that a CA issues. This online verification process requires additional information exchange during a transaction and is not suitable for restricted connectivity scenarios due to the communication restrictions of such scenarios.
- *Implicit Verifiable Certificate:* Unlike the previous case, here the authenticity of the public key is established during the use of the keys. This concept was introduced in 1991 [51] and is known as *self-certified public keys*. In this scheme, the user's public key is derived from the signature of his/her secret key with his/her identity, and the SA signs the key using the secret key of the system. Thus, public key authentication can be achieved implicitly with signature verification and without the use of a CA-issued certificate.

  Schemes based on self-certified public keys are an alternative to using asymmetric cryptography in restricted connectivity scenarios due to the elimination of the communication with the CA for certificate verification. Some examples demonstrating the applicability of self-certified public keys in the MPS are the proposals Téllez et al. presented in recent years, which were discussed in Chap. 3.

Based on the self-certified public key system that [51] proposed, [181] suggested two new schemes. The first one is an authenticated encryption scheme that only allows a specified receiver to verify and recover the message. The second (known as the authenticated encryption scheme with message linkages) allows a large message to be divided into a sequence of message blocks where each message block is signed individually. This scheme sends long messages more efficiently than the first scheme.

### 4.2.6 *Security Vulnerabilities, Threats, Risks, and Protection Solutions*

Since security vulnerabilities of communication technologies (such as GSM, Bluetooth, RFID, etc.) are usually ignored while analyzing the security aspects of an m-payment system, it is important that MPS designers take a holistic view when performing a security analysis of mobile payment systems during their design and implementation [3]. In general, a secure MPS has to meet the transaction security properties discussed in Sect. 4.1 to counter potential threats.

Table 4.1 presents a summary of the types of vulnerabilities and threats and their corresponding risks in an MPS environment together with relevant protection solutions.

## 4.3 Security and Constraints of Mobile Payment Systems

The literature highlights several reasons to explain why MPSs are not yet secure. However, the two main reasons are the constraints of wireless environments and the security of the MPSs [92]. Both aspects are discussed in the following sub-sections.

### *4.3.1 Constraint of Wireless Environments*

The features of wireless networks and resource constraints of mobile devices are aspects that impact the execution of payment transactions in wireless environments [58, 72, 150]. Therefore, these aspects should be considered in the design of new MPSs or when payment systems designed for fixed networks must be adapted for the wireless environment.

#### 4.3.1.1 Resource Limitations of Mobile Devices

Despite the wide range of mobile devices available in the market, they have common limitations, including the following:

- Their processor has less computational capability compared with PCs.
- They work with battery, and, due to the short range of batteries, they can only operate for periods of time shorter than those of PCs.
- They have limited storage space that affects, among other things, the type of cryptographic algorithms that can be applied to these devices.

Mobile devices with the limitations mentioned above have difficulties performing efficiently, cryptographic operations with high computational requirements such as public key operations. Additionally, because mobile devices have low computational capabilities, they require more time to conclude a payment transaction a PC does, due to its higher processing capabilities. Moreover, the public key cryptographic operations, often used on fixed network devices (such as PCs), require a certificate verification process and storage space for each public key certificate on the mobile device (which has limited space).

**Table 4.1** Vulnerabilities, threats, and risks in MPSs along with corresponding protection solutions [3, 76, 79]

| Vulnerability | Threat | Risk | Protection solutions |
|---|---|---|---|
| Over The Air (OTA) transmission between phone and Point Of Sale (POS) | Interception of traffic | Identity theft, information disclosure, replay attacks | Trusted Platform Module (TPM), secure protocols, encryption |
| Inadvertent installation of malicious software (such as malware, spyware, etc.) on mobile phone by user | Downloaded application intercept of authentication data | Theft of authentication parameters, information disclosure, transaction repudiation | Authentication of both user (PIN) and application (digital signature by trusted third party); TPM; anti-malware/anti-spyware/anti-virus |
| Absence of two-factor authentication | User masquerading | Fraudulent transactions, provider liabilities | Two-factor authentication |
| Changing or replacing mobile phone | Configuration and setup complexity | Reduced adoption of the technology; "security by obscurity" | Simplified user interface, security parameters in TPM set by trusted party |
| Smartphone Internet and geolocation capabilities | Malware on mobile device; poor data protection controls at merchant/payment processor | Data disclosure and privacy infringement; profiling of user behavior | User control of geolocation features, cryptographically supported privacy, trusted platform module, vetted authorization and accounting |
| POS devices are installed at merchant premises | Masquerade attacks; tampering with POS | Theft of service, replay, message modification | POS vendor vetting, message authenticators, vetted authorization and accounting |
| Lack of Digital Rights Management (DRM) on mobile device | Mobile device user illegally distributes content (e.g., ringtone, video, games) | Theft of content, digital piracy, risk to provider for digital rights infringement, loss of revenue to content provider or merchant | DRM incorporated in smartphone TPM design, cryptographically supported DRM |

| | | | |
|---|---|---|---|
| Weakness of Global System for Mobile Communication (GSM) encryption for OTA transmission; SMS data in cleartext on mobile network | Message modification, replay of transactions, evasion of fraud controls | Theft of service or content, loss of revenue, illegal transfer of funds | Strong cryptographic protocols, SMS message authenticators, encryption |
| Weakness of GSM protocol for mutual authentication | Impersonation or man-in-the-middle attacks | Eavesdrop, modify, delete, re-order, replay, and spoof signaling and user data messages exchanged between the two parties | Mutual entity authentication capable of providing secure bilateral authentication |
| Weakness of encryption technique on Subscriber Identity Module (SIM) cards | Tampering or cloning of SIM Card | Carry out malicious transactions on behalf of the user; theft of stored payment credentials | SMS firewall on the phone; state-of-art cryptography with sufficiently long keys; PIN for the SIM card |
| Weak cryptographic keys or predictable random numbers cryptographic APIs provided by development platforms | Cryptanalysis and dictionary attacks | Unauthorized access to restricted functionalities | Secure cryptography Application Programming Interfaces (APIs) provided by third party with cryptographic algorithms capable of generating truly unpredictable random numbers and strong cryptographic keys |

Despite the worldwide growth of smartphone (mobile devices with high computational capability) sales in recent years, they do not reach half of the mobile device market according to the IMS Research firm [18]. Therefore, a huge number of WAP-enabled phones with the previously mentioned constraints still exist in the market.

### *4.3.2 Characteristics of Wireless Networks*

Wireless network refers to any type of computer network where the nodes do not need a physical connection (cable) to communicate. This ability to communicate wirelessly constitutes one of the main advantages of these networks over wired networks, but it also introduces new security issues that should be considered to prevent the theft of information and access of intruders. Wireless networks' characteristics include the following:

- The bandwidth of wireless networks is lower than fixed networks.
- Network connections in this type of network are less reliable because there are frequent packet losses, which are higher than in fixed networks. Hence, it is necessary to retransmit the packets, which can cause a high latency.
- Compared to fixed networks, the wireless networks have a higher connection cost.
- Outsiders can easily eavesdrop on data transmitted over wireless networks.

From the characteristics discussed above, one can conclude that the execution of payment transactions in wireless networks is time consuming mainly due to the poor performance of these types of networks. These characteristics in combination with the low computational capabilities of mobile devices increase the time required to complete each payment transaction.

Furthermore, because the connection cost of communication in fixed networks is lower than that of wireless networks, execution of payment transactions in wireless networks through mobile devices involves additional charges for users. Additionally, since outsiders can easily eavesdrop on the data transmitted over wireless networks, the devices should use highly secure cryptographic techniques to prevent such breaches of privacy, but the operations require devices to have high computational capabilities and high-speed wireless networks, which results in high costs for users.

# Chapter 5
# Future Challenges and Opportunities

In the last couple of years, mobile devices have become for the millennials especially an instrument not only for communication but also for coordinating their social life. Furthermore, the young generation uses mobile devices to conduct electronic transactions instead of ordinary PCs. Therefore, m-commerce, a synonym for wireless e-commerce, supports electronic commerce capabilities via various types of wireless mobile devices and provides customers with access to retail outlets anywhere, anytime, from any device. However, although m-commerce is beneficial for consumers who keep a mobile phone by their side all the times, the technology also allows ubiquitous access to any consumer regardless of where he/she is. There are many factors that could hinder the success of m-commerce applications. A secure environment is necessary to make m-commerce ubiquitous and widely accepted, especially for those transactions involving money because consumers need to feel secure to conduct their electronic transactions. Depending on the point of view of the different participants in a mobile commerce scenario, the following security challenges have been previously identified [158]:

- *The Mobile Device:* Confidential user data on the mobile device as well as the device itself should be protected from unauthorized use by employing security mechanisms that include: user authentication, secure storage of confidential data, and security of the OS.
- *The Radio Interface:* Access to a telecommunication network requires the protection of transmitted data in terms of confidentiality, integrity, and authenticity. In particular, the user's personal data should be protected from eavesdropping.
- *The Network Operator Infrastructure:* As the security mechanisms for the end user frequently terminate in the access network, it is necessary to handle the security of the user's data carefully within and beyond the access network. Moreover, since the user receives certain services for which he/she has to pay, both the network operator and the user must be assured of correct charging and billing.



J. Téllez, S. Zeadally, *Mobile Payment Systems*, Computer Communications and Networks, DOI 10.1007/978-3-319-23033-7_5

- *The Kind of M-commerce Application:* The security of m-commerce applications that involve payment is required in order to reassure customers, merchants, and network operators. The authenticity, confidentiality, and integrity of transmitted payment information is important. Also, non-repudiation is important as well as the confidence of the customer about the delivery of goods or services.

Mobile commerce faces its own unique challenges, which are more directly related to its set of unique features, such as m-commerce's anytime, anywhere, always available, and immediate connectivity, and the smartphone's tracking ability to reveal more personal related information (such as a user's location, device serial number, international mobile equipment identity (IMEI), integrated circuit card identifier (ICCID) or SIM card ID, social relationships, life style, preferences, or behavior patterns). In this context, the protection of consumers' information privacy is an important challenge. Furthermore, for m-commerce marketers, vendors, and policy developers, a study about consumers' demographic differences is necessary to minimize their concerns for information privacy [196]. The aforementioned challenges associated with m-commerce can affect the healthy growth of mobile commerce, and therefore future research is required in order to address them. Moreover, in the context of mobile commerce, several other issues such as the impact of location-based digital couponing and context-aware advertisement on information privacy needs to be explored further.

Though MPSs offer various benefits, they are affected by the mobile commerce issues as we explained in the previous paragraphs and by several challenges introduced by them, which must be overcome in order for a successful implementation of mobile payment to be widely accepted and adopted as a mode of payment. In the near future, the research community should address the challenges of MPSs, which are mostly concerned with the following aspects:

(a) The security of emerging technologies, which are now increasingly integrated into mobile devices and are converging with existing ones [48].
(b) The security of payment scenarios (some of them with communication restrictions) where the payment could occur.
(c) Improved performance of payment transactions by using alternative cryptography techniques (such as elliptic curves) to reduce processing overheads.
(d) The design of a regulatory framework and the development of widely accepted standards for the development of mobile payment applications.

Users face several challenges in MPSs, which suffer from a lack of payment options around their mental model development, usability issues, pre-purchase anxiety, and trust issues, ease-of-use, and support for routine purchases with mobile payments. Hence, it is necessary to design MPSs that can provide a better mobile payment experience, which incorporates users' routines and behaviors and trust mechanisms. A brief discussion of these follows [66].

- *New Users' Routines and a Lack of Benefits:* For new users, the habit of using mobile payments is still forming. Also, despite the fact that users understand how to use them, many do not see the benefits of using mobile payments instead of a credit card when the mobile payment service offers the same benefit, that is, payment without cash. Moreover, as mobile payments are currently only available in a small number of situations and stores, practices of new users will not change unless more stores adopt a mobile payment service option. Therefore, significant societal shifts in payment options and usage are required in order to motivate users to change their routines and use mobile payment services in the near future.

- *Mental Model Development and Usability Issues:* Mental models often help to shape behavior and explain a person's thought process on how something works. Some users cannot understand the concept of paying with their mobile device or how to start the process. Other users do not have the "mental model" to see their mobile device as a payment source. A common theme across mobile payment services is the lack of visual or audio indicators for feedback around transactional information. This is important for users because they must feel satisfied after a transaction is complete.
- *Pre-purchasing Anxiety:* Before making a purchase, a user must prepare his/her mobile device for use, but may become nervous that the mobile device might not be ready to be used/scanned when it is their turn in line. This might occur because, for example, the mobile device was turned into screen saver mode and a password needs to be entered, or the barcode is not ready to be scanned. This situation creates a pre-purchase anxiety in some users when using mobile payments.
- *Trust Concerns:* There are several trust concerns that are related to mobile payments, which include access to information and fragmented payment solutions. Users can have serious issues with information access across a variety of MPSs or have trust concerns over the security of paying through a barcode displayed on their mobile devices. Moreover, another worry is around the security of personal information that a user has to send over Wi-Fi or other networks. Similarly, the possibility that data displayed on the screen of the user's mobile device may be visible to outsiders is another issue around access to information. Regarding the fragmented payment solutions, users are concerned about leaving their money or personal information untouched even if they are not used. Therefore, a single global account is desirable in order to simplify the handling of the user's money and keep track of account charges in order to ease trust concerns and overall fear of money loss.

Mobile devices play an important role in MPSs and each one has certain characteristics that influence its usability and the services that the end user can receive, such as limited display, memory and CPU processing power, supported operating systems, availability of internal smartcard reader, network connectivity (bandwidth capacity), input device (availability of keyboard and mouse), short-range wireless communication (NFC, RFID), and biometric capabilities. In addition, mobile devices' security is still an open issue, because they are exposed to the same attacks as desktop computers and often with poorer security. Therefore, mobile device security has to deal with the protection of the systems and information from unauthorized access, disclosure, use, disruption, recording, or destruction. The security challenges that mobile devices face include:

(a) The inability of users to make good security choices when an application requests access to gain the necessary permissions to protected resources.
(b) Mobile web browsers can lead users to malicious websites that can inject JavaScript and other executable code that automatically downloads malware, provides a channel for remote exploits, or hijacks the system and turns it into a zombie [169], all through the browser.
(c) Root exploits are another way that mobile malware infects devices in order to give a user or application superuser privilege. This enables the attacker to install all sorts of applications (many of which are either malware or have been infected by malware) to gain full access to the device remotely without user detection.
(d) SMS-based attack with the primary objective of transmitting premium-rate SMS messages to offshore numbers and sending mass SMS spam advertisements for fraudulent products and services.

To address the above challenges, various approaches and solutions have been proposed. However, some solutions require the presence of an on-device, lightweight agent, with sophisticated policy decision making whilst others focus on the core mobile OS or the hypervisor. As for the challenge of an on-device solution, the software solution must support a large number of operating system platforms and versions (due to the diverse mobile operating systems), which impose additional restrictions on mobile applications. The other challenge is resource constraints such as the computation complexity of the on-device software solution that should not use a lot of CPU time because its impact can be high enough to make the device unusable [102].

Recently, to meet the mobility demands of today's employees, organizations have started to adopt the concept of Bring Your Own Device (BYOD). BYOD is an emerging business trend where companies allow employee-owned technologies in the workplace to improve productivity gains and cost benefits by enabling employees to use their technology of choice in the workplace. However, this new technology is accompanied by new security challenges that must be addressed. Three of the most frequently cited BYOD risks include [32]:

(a) *Data Leakage:* This is related to the threat to organizations of intended or inadvertent disclosure of sensitive data (such as private customer information and proprietary company information) stored on employee-owned devices.
(b) *Loss of Control and Visibility:* This refers to the significant security risks introduced by BYOD for organizations because they no longer control the device in which the company data is stored. Therefore, the enforcement of security policies becomes difficult for issues such as data leakage, theft, and regulatory compliance. As organizations increasingly lose control over the security of their terminal devices, employees have a more critical role to play in upholding organizational security and must be incorporated into the security model of the organization. Also, managing the BYOD devices leads to other

technical challenges related to incorporation of device-based authentication to approve/deny access, the possibility that personal devices can infect the company network with malware, and the constant technical change of the various OSs of the mobile handsets, which make them outdated very quickly.

(c) *Ease of Device Loss:* This refers to the exposure of BYOD devices and the information stored on them if the device is lost or stolen due to the relatively small size and portability of mobile devices.

Several proposals of MPSs described in the literature to allow clients to pay for goods and/or services online from a mobile device are designed to accept only one payment method per transaction. Therefore, these approaches do not contemplate those situations in which the client may want to combine more than one payment method in a transaction. This limitation constitutes a challenge that must be addressed by MPSs designers in future designs in order to allow clients to use two or more payment methods within the same order.

In mobile payments, authentication is a critical concern in the mobile wireless environment. Usually, identification and authentication methods can be broadly classified into three main categories: knowledge-based authentication (e.g., passwords, PINs, and so on.), token-based authentication (e.g., smart card, USB dongle, etc.), and biometric authentication (e.g., fingerprints, iris scans, etc.), which is considered to be the next generation authentication technique. Biometric techniques use physiological or behavioral characteristics to determine or verify identity. Typical biometrics technologies include fingerprint, finger vein, face, palm print, and gait, to achieve authentication in a more trustworthy manner.

Nowadays, biometrics-embedded mobile devices are becoming increasingly popular and their use in mobile payments has helped the development of more secure biometric-based MPSs. However, there are challenges that still need to be addressed about the use of biometrics for securing mobile payments. The security of biometric templates has a high priority that needs more attention because the number of an individual's biometric sources is limited (such as ten fingers, one face, and two eyes) and cannot be reset like passwords or reissued like tokens. Therefore, a biometric template that is compromised can lead to a permanently compromised biometric ID, which is hard to replace. In a biometric authentication system it is impossible to find two perfectly matched feature sets because they tend to be different every time under different sensing and ambient conditions and various interactions between the user and the sensor. As a result, the recognition error rates of a biometric-based payment systems depend on the choice of biometric traits and the quality of the extracted biometric feature data. Most fingerprint verification techniques rely on accurate registration, which is still very challenging especially for poor quality images [15].

Due to the limited processing capability of mobile devices, mainstream mobile manufacturers are still unable to apply advanced security protocol such as the public key cryptosystem based on RSA algorithm to mobile devices. Several past efforts in recent years have proposed efficient and practical methods to implement the RSA algorithm running on the Texas Instruments TMS320C55x family [69].

Other studies have been exploring the efficiency on modern mobile devices of variants of the RSA cryptosystem (such as CRT, Multi-Prime RSA, Multi-Power RSA, Rebalanced RSA, R-Prime RSA, and Hwangs method). However, these approaches are highly theoretical studies and proposals. To better understand their impact in real-life applications, their effectiveness must be evaluated in a real networking environment. The results obtained will help MPSs designers to determine the appropriateness of using the RSA algorithm in their systems.

An alternative public key scheme for those MPSs that have communication restrictions (where an engaging party is not able to communicate with a CA to validate a certificate during a transaction) is the self-certified public key scheme, where the user's public key is derived from the signature of his/her secret key along with his/her identity, signed by the SA using the secret key of the system. However, a review of the literature shows that self-certified public key schemes suffer from the following open problems that remains to be solved in the future: (a) the expiration of the kind of certificate used in these schemes is not defined, and (b) the accountability for the actions of the originator and the recipient cannot be established because the following non-repudiation services are not supported: Non-Repudiation of Origin (NRO), Non-Repudiation of Receipt (NRR), Non-repudiation of Submission (NRS), and Non-repudiation of Delivery (NRD). There is also no implementation of the self-certified public key scheme in mobile devices that demonstrates if its performance is suitable for these types of devices.

ECC is an alternative approach to public key cryptography that relies on the elliptic curve algorithm, which reduces dramatically the key size necessary to accomplish the same level of security provided in conventional public key cryptographic schemes. Thus, ECC can offer the following properties, which are helpful for implementing encryption on resource-constrained mobile devices: faster computations, lower power consumption and memory, and bandwidth savings. ECC algorithms offer an interesting approach with many benefits for existing or new MPSs. ECC algorithms offer an interesting approach with many benefits for existing or new MPSs. However, it is still necessary to implement them to analyze their performance on mobile devices in order to determine if they are suitable for ensuring secure payment protocols. Moreover, this will enable ECC algorithms to gain wide acceptance among MPSs designers who should explore the possibility of incorporating them into their new MPSs [76].

An NFC-based mobile wallet system is a type of mobile wallet that requires a special chip and an application (app) on the smartphone to allow the user to login and make a payment. This type of wallet requires the consumer to log in to the app, using perhaps a secure PIN or password. Once the user has logged in, he/she can use the app and scan his/her phone on a nearby POS terminal to send the payment request to the merchant. From the consumer and merchant point of view, this is the same procedure as EMV[1] PIN payments. The main benefit of this payment method for the consumer is the convenience that it brings in busy environments such as train stations.

[1]EMV is a technical standard for smart payments and for payment terminals and automated teller machines.

Despite the benefits offered by NFC-based mobile wallet systems, the NFC technology is not without flaws. Currently, there are few mobile phones on the market equipped with NFC chips and few merchants have NFC readers. This is delaying the growth of this kind of mobile wallet payments but NFC still continues to represent the most promising path toward widespread mobile payment adoption. In addition, there are others obstacles to overcome before this becomes the technology of choice for mobile payments. These obstacles include: consumers and merchants are comfortable with the current POS environment because they understand it and they do not need to be trained on its use; it has higher infrastructure requirements than some other forms of mobile wallets; back-end systems need to execute traditional batch processes all in real time and to interact with the embedded secure element; and TSMs are required to download card credentials securely into the phone over-the-air and manage those credentials on an ongoing basis. These obstacles introduce insecurity into the mind of the consumer [28, 143].

The Mobile Cloud Computing (MCC) term was introduced after the concept of Cloud Computing and refers to an infrastructure where both the data storage and the data processing occur outside the mobile device. Mobile cloud applications move the computation power and storage away from the mobile device to the cloud [35]. Various mobile applications have taken advantage of MCC as is the case of MPSs. A cloud-based MPS is a type of proximity payment that stores payment credentials (used to authenticate the payment transaction) on a remote server, rather than at the mobile device. Notwithstanding the advantages offered by cloud-based MPSs, the following challenges still need to be addressed in the future [29, 76]:

- Merchants must implement additional infrastructure to accept cloud payments at the POS, and the customer must register with each individual merchant before making a payment.
- Some cloud-based transactions may be treated as Card-Not-Present (CNP), which refers to a payment card transaction made where the cardholder cannot physically present the card for a merchant's visual examination at the time of the transaction. This results in higher transaction fees for the merchant.
- As cloud payments require Internet connectivity, a transaction may not work or be interrupted due to connectivity issues, particularly if access to the cloud fails and there are no back-up payment credentials stored on the mobile phone.
- A payment transaction may require more time because transmission to the cloud is slower than NFC to POS.
- Some security issues remain unsolved and these include: (a) payment data and stored payment credentials in the cloud could be compromised if the cloud server is attacked; (b) payment data should not be transmitted through SMS or email due to the absence of data encryption on cloud platforms [29]; and (c) data privacy remains a key concern for payment data stored in the cloud because of cloud owners.
- Data privacy is a concern for payment data stored in the cloud because consumer data is controlled by cloud providers who have both a legal and ethical responsibility to protect it. Therefore, they need to make sure that their underlying

payment services are secure and resilient, and the consumer information will not be shared with other businesses. They must also ensure that they do not or make mining results of consumers' data available to companies interested in monitoring consumer spending behaviors without the explicit permission of the consumer.

There are a number of handsets on the market today that have a secure element (called SE and refers to a tamper-resistant platform capable of securely hosting applications and their confidential and cryptographic data in accordance with the rules and security requirements set forth by a set of well-identified trusted authorities). Consequently, it is necessary to consider those cases where a mobile phone has NFC capabilities but no secure element that can be easily accessed by financial institutions. Therefore, we require a payment solution where the consumer's credentials are stored somewhere in the cloud instead of a secure element locally on the mobile phone and delivered from the cloud to the mobile phone at the time of a purchase because there is no secure element that is tightly controlled by a MNO or other entity that is restricting access. Therefore, the credentials could be transmitted to a virtual card in the phone as needed and delivered to the POS reader via the NFC connection. Once the credentials have been received by the mobile phone, they are securely transmitted to the point of sale through the NFC connection. This approach, also called NFC cloud-based wallet, uses cloud computing to manage NFC payment applications. This payment method results in flexible and secure management, personalization, ownership of the applications and instant fraud detection because the system runs in an online mode. However, the practice of storing payment credentials in the cloud has not been thoroughly tested as a mobile wallet solution. In addition, a high-quality Internet connection (via WiFi or 3G/4G/5G) is required to deliver credentials from the cloud to the phone at the POS. New approaches are still required for cloud-based NFC data transmissions to minimize the security risks that are involved during a transaction process. Moreover, the payment card industry should collaborate with international payment standards to develop additional security standards (such as PCI DSS [121]) in order to protect sensitive information [143].

Today, mobile devices equipped with cameras can be used with barcodes to perform various functions, including making mobile payments, accessing sites on the Internet, downloading products, searching reviews and information about products, or accessing loyalty programs. Nevertheless, it is not easy for many users to distinguish between QR codes from trusted and untrusted sources. Therefore, scanning a QR code is not a safe practice because users need to first decode the QR code in order to decide whether the content can be trusted because it is not human-readable. In addition, malicious QR codes can contain URLs with hidden malware, or redirect the user to a cloned website to commit fraud, download malware, or phish for credentials. Hence, these security risks need more attention from MPS designers who should include countermeasures to thwart the attacks in MPSs based on QR code, such as adding passcode protection to prevent use of the system if the phone is lost or stolen, or use of public certificates for mutual authentication between the customer and the merchant for MPSs based on restricted connectivity scenarios [29, 87, 89].

Krombholz et al. [89] have identified some of the major research challenges that need to be addressed in order to improve QR code security with an emphasis on usability and security aspects, which are briefly described below:

- **Security Awareness Challenges:** Internet users perceive privacy challenges and security vulnerabilities very differently because they have different concerns regarding security vulnerabilities when interacting with QR codes.
- **Usable Security Design Guidelines:** Since most users cannot successfully detect phishing attacks, it is necessary to do a more detailed analysis on security, privacy, and usability factors when designing a software that supports the user's decision-making process about the trustworthiness of a URL. Another important challenge is the development of a secure and usable multi-layer framework for QR code processing, as well as the recommendation of design guidelines to harden the QR code itself and the reader software, and supporting the user to detect potential threats.

Restricted connectivity scenarios opens tremendous business opportunities and research challenges because they consider situations in which two entities of the restricted connectivity model cannot communicate with each other directly and must do so through a third party. In this context, security is an important key aspect that designers of MPSs must take into consideration in order to provide the same security levels as those based on the full connectivity scenario to generate the necessary confidence to stimulate the creation of MPSs on this type of scenarios (often encountered in the real world).

Although a few m-payment solutions have been proposed for the restricted connectivity scenarios discussed earlier, all of them rely on online payment protocols. One approach is to design new m-payment solutions based on others existing mobile payment methods such as: micro-payments, pre-paid, token-based, and so on. Also, the restricted connectivity scenario shown in Fig. 5.1 (called Disconnected Model) appears to remain unexplored by the research community as no proposal could be found in the literature. More research is needed to investigate the design and architectural issues of MPSs based on the disconnected restricted scenario.

Several online payment protocols, *kiosk-centric*, *client-centric*, and *PG-centric*, for the restricted connectivity scenarios [70–74, 77, 78, 104, 159, 174] have been presented in the literature. These payment protocols employ different cryptographic schemes such as symmetric cryptography, self-certified public keys, and self-certified public keys based on elliptic curve. Several of them were designed for mobile environments and others for VANETs. Moreover, some of these payment protocols have also been implemented in real networking environments to measure their actual performance.

VANETs are envisioned to support the development of a wide range of attractive applications, such as payment services, which require the design of payment systems that satisfy additional requirements associated with VANETs. Payment systems for VANETs must deal with performance and security (mainly with the prevention of attacks made by malicious nodes) issues when they are developed in order to allow payments among nearby vehicles using various routing protocols. In

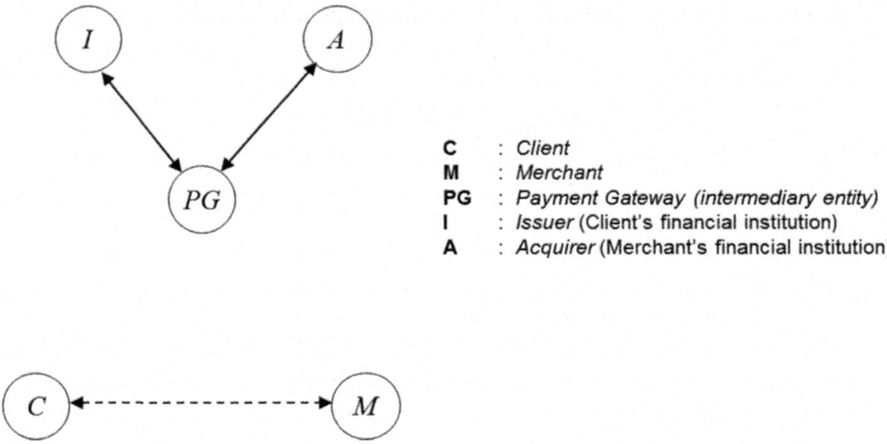


**Fig. 5.1** Scenario with client and merchant disconnected from the PG: disconnected model

addition, use of a stored-value card (a kind of smartcard that can store monetary value that is used in a wide range of applications in micro-payment environment) is a promising option to provide an added-value service of payment protocols for VANET. However, this needs more research in the future to address issues like security risks, unrestricted anonymity, and use in restricted connectivity scenarios [24, 55, 73, 77, 194].

Vehicular Cloud Computing (VCC) combines the benefits of cloud computing to serve the drivers of VANETs with a pay-as-you-go model [17, 189]. VCC opens up business opportunities to develop new payment systems for the vehicular cloud computing environment. However, vehicular clouds have several security and privacy challenges, which must be addressed by payment system designers in order to create payment systems based on vehicular clouds. These challenges include [189]: (a) verifying the authentication of users and the integrity of messages due to the high mobility of nodes, (b) ensuring the confidentiality of sensitive messages by using the cryptographic algorithm, (c) providing data isolation to ensure the security of stored data on the cloud, and (d) securing data access to protect stored data on the cloud against unauthorized access.

The 5G mobile communications technology is the next generation of the existing 4G Long Term Evolution (LTE) network technology, which will allow users to transmit large data files including high quality digital movies practically without limitation. Additionally, 5G will enhance the latency, battery consumption, cost, and reliability that will diminish the cost of communications over wireless networks when executing payment transactions and help to support real-time applications such as voice over IP (VoIP), gaming, and videoconferencing. Nevertheless, the security solutions used by distinct wireless, mobile, and cellular networks makes end-to-end security solutions an important challenge that still needs to be addressed to support future secure MPSs and applications [76]. Some of the threats that future research should focus on in order mitigate them include: (a) third parties masquerading as legitimate users resulting in theft of service and billing frauds, (b) spoofing attacks that can lead to misdirected communication and Internet banking related frauds, (c) the Spam over Internet telephony (SPIT) (which refers to the undesired, automated, pre-recorded, bulk telephone calls made using Voice over Internet Protocol (VoIP)), the new spam over VoIP, and (d) compromising the operations of a single operator, which can disrupt the operation of the entire network infrastructure because all the network operators and service providers would share a common core network infrastructure.

New concerns about the security of payment systems that must support electronic payment transactions performed over wireless networks has been raised because they have to offer the same level of security as those designed to support electronic payment transactions on fixed networks. However, the execution of payment transactions in wireless environments suffers from a number of limitations. To address these constraints, designers of wireless payment systems must further explore innovative, robust, cost-effective, secure mobile payment protocols. One possible solution to reduce the computational and communication requirements of mobile users is to replace the high computational cryptographic operations with other cryptography techniques that incur lower computational and communication overheads (e.g., replacing public key cryptography operations with symmetric key operations and hash functions or by using ECC). However, since different types of cryptographic operations provide different levels of security, there is a risk that the payment system may not satisfy all the security requirements needed by the underlying MPS (as in the case of symmetric cryptography operations where, although they require less computational power than the public-key operations, they do not provide the non-repudiation property). Future MPSs need to carefully evaluate the trade-off between transaction performance and security [76] during their designs and implementations.

Several research studies on m-payments have been reported in the literature. These efforts proposed solutions to problems found in the past and also identified open research issues and challenges for m-payment systems that need to be addressed by researchers. A study conducted by Aoki et al. [8], argued that the developers of technology and the users of it have a symbiotic relationship whereby each responds to changes made by the other. In a similar vein, as m-payment systems and their technologies continue to emerge and evolve, it is imperative that academic research investigates users' adoption related to emerging m-payment technologies. Moreover, as current research has identified multiple complex and interrelated factors that affect the adoption of m-payment systems, further research is needed to develop a more attractive m-payment model for end-users. Another important future area of research is the investigation of the various factors that determine the adoption of m-payment systems by users. Furthermore, more merchant centric research is also needed because, in the past literature, we found a much higher volume of research conducted from the consumers' perspective rather than the merchants' or m-payment service providers' perspective [165].

# Bibliography


1. J.L. Abad-peiro, N. Asokan, M. Steiner, M. Waidner, Designing a generic payment service. IBM Syst. J. **37**(1), 72–88 (1997)
2. R. Abbadasari, R. Mukkamala, V. Valli Kumari, Mobicoin: digital cash for m-commerce. in *First International Conference Distributed Computing and Internet Technology (ICDCIT 2004)* (2004), pp. 441–451
3. S. Agarwal, M. Khapra, B. Menezes, N. Uchat, Security issues in mobile payment systems, in *5th International Conference on E-Governance* (2007), pp. 142–152
4. A. Ali, A study of security in wireless and mobile payments. Master's thesis, Department of Electrical Engineering, Linköping University (2010)
5. M.V. Alizade, R.A. Moghaddam, S. Momenebellah, New mobile payment protocol: mobile pay center protocol (MPCP), in *3rd International Conference on Electronics Computer Technology (ICECT 2011)* (2011), pp. 74–78
6. M. Al-Meaither, Secure electronic payments for Islamic finance. Ph.D. thesis, Royal Holloway, University of London (2004)
7. L. Antovski, M. Gusev, M-payments, in *25th International Conference on Information Technology lnferfaces (ITI 2003)* (2003)
8. K. Aoki, E.J. Downes, An analysis of young people's use of and attitudes toward cell phones. Telematics Inform. **20**(4), 349–364 (2003)
9. N. Asokan, P.A. Janson, M. Steiner, M. Waidner, The state of the art in electronic payment systems. IEEE Comput. **30**(9), 28–35 (1997)
10. S. Bakhtiari, A. Baraani, M.-R. Khayyambashi, Mobicash: a new anonymous mobile payment system implemented by elliptic curve cryptography, in *World Congress on Computer Science and Information Engineering (WRI 2009)* (2009), pp. 286–290
11. C. Bangdao, A.W. Roscoe, Mobile electronic identity: securing payment on mobile phones, in *5th IFIP WG 11.2 International Workshop (WISTP 2011)* (2011), pp. 22–37
12. M. Becher, F.C. Freiling, Towards dynamic malware analysis to increase mobile device security, in *Sicherheit 2008* (2008), pp. 423–433
13. M. Bellare, J.A. Garay, R. Hauser, A. Herzberg, H. Krawczyk, M. Steiner, G. Tsudik, E. Van Herreweghen, M. Waidner, Design, implementation and deployment of the *i*KP secure electronic payment system. IEEE J. Sel. Areas Commun. **18**(4), 611–627 (2000)
14. M. Bellare, J.A. Garay, R. Hauser, A. Herzberg, H. Krawczyk, M. Steiner, G. Tsudik, M. Waidner, *i*KP – a family of secure electronic payment protocols, in *First USENIX Workshop on Electronic Commerce* (1995)


 
J. Téllez, S. Zeadally, *Mobile Payment Systems*, Computer Communications and Networks, DOI 10.1007/978-3-319-23033-7

15. N. Ben-Asher, H. Sieger, A. Ben-Oved, N. Kirschnick, J. Meyer, S. Möller, On the need for different security methods on mobile phones, in *Proceedings of the 13th International Conference on Human Computer Interaction with Mobile Devices and Services (MobileHCI 2011)* (2011)
16. J. Bickford, R. O'Hare, A. Baliga, V. Ganapathy, L. Iftode, Rootkits on smart phones: attacks, implications and opportunities, in *Proceedings of the Eleventh Workshop on Mobile Computing Systems & Applications (HotMobile 2010)* (2010), pp. 49–54
17. S. Bitam, A. Mellouk, S. Zeadally, VANET-cloud: a generic cloud computing model for vehicular ad-hoc networks. IEEE Wirel. Commun. Mag. **2**(1), 96–102 (2015)
18. M. Brownlow, Smartphone statistics and market share (2012), http://www.email-marketing-reports.com/wireless-mobile/smartphone-statistics.htm#smartphones. Last accessed Dec 2012
19. R. Byler, What is money? in *2nd Annual Conference on Mid–South College Computing (MSCCC 2004)* (2004), pp. 200–209
20. B. Campbell, J. Rosenberg, H. Schulzrinne, D. Gurle, C. Huitema, Session initiation protocol (SIP) extension for instant messaging. RFC 3428, IETF (2002)
21. M. Carr, Mobile payment systems and services: an introduction. Mobile Payment Forum (2007)
22. S. Chari, P. Kermani, S. Smith, L. Tassiulas, Security issues in m-commerce: a usage-based taxonomy, in *E-Commerce Agents, Marketplace Solutions, Security Issues, and Supply and Demand* (2001), pp. 264–282
23. K. Chaudhary, X. Dai, P2p-netpay: an off-line micro-payment system for content sharing in p2p-networks. J. Emerg. Technol. Web Intell. **1**(1), 46–54 (2009)
24. C.-L. Chen, W.-C. Tsai, Using a stored-value card to provide an added-value service of payment protocol in vanet, in *Seventh International Conference on Innovative Mobile and Internet Services in Ubiquitous Computing (IMIS 2013)* (2013), pp. 660–665
25. W.-D. Chen, K.E. Mayes, Y.-H. Lien, J.-H. Chiu, NFC mobile payment with citizen digital certificate, in *The 2nd International Conference on Next Generation Information Technology (ICNIT 2011)* (2011), pp. 120–126
26. W. Chou, L. Washington, Elliptic curve cryptography and its applications to mobile devices, in *IEEE INFOCOM 2004* (2004)
27. J. Claessens, Analysis and design of an advanced infrastructure for secure and anonymous electronic payment systems on the Internet. Ph.D. thesis, Katholieke Universiteit Leuven (2002)
28. C. Cox, S. Sanchez, Transforming the customer experience: the promise of mobile wallets. Technical report, First Data Thought Leadership (2013), http://tex.stackexchange.com/questions/5957/bibtex-entry-for-white-papers-and-technical-reports
29. M. Crowe, E. Tavilla, Mobile phone technology: "smarter" than we thought (2012), http://www.bostonfed.org/bankinfo/payment-strategies/index.htm. Last accessed 30 Sept 2013
30. E.C. Cuervo, Implementación de un monedero digital móvil. Master's thesis, Centro de Investigación y de Estudios Avanzados del Instituto Politécnico Nacional (Cinvestav) (2005)
31. X. Dai, J. Grundy, Netpay: an off-line, decentralized micro-payment system for thin-client applications. Electron. Commer. Res. Appl. **6**(1), 91–101 (2007)
32. A. Dedeche, F. Liu, M. Le, S. Lajami, Emergent BYOD security challenges and mitigation strategy (2013), https://minerva-access.unimelb.edu.au/bitstream/handle/11343/33340/300297_2013_Lajami_Strategy.pdf?sequence=1. Last accessed Aug 2014
33. G. Delac, M. Silic, J. Krolo, Emerging security threats for mobile platforms, in *34th International Convention on Information and Communication Technology, Electronics and Microelectronics (MIPRO 2011)* (2011), pp. 1468–1473
34. N. Delic, A. Vukasinovic, Mobile payment solution-symbiosis between banks, application service providers and mobile networkoperators, in *Third International Conference on Information Technology: New Generations (ITNG 2006)* (2006), pp. 346–350
35. H.T. Dinh, C. Lee, D. Niyato, P. Wang, A survey of mobile cloud computing: architecture, applications, and approaches. Wirel. Commun. Mob. Comput. **13**(18), 1587–1611 (2013)

36. S. Dominikus, M.J. Aigner, mcoupons: an application for near field communication (NFC), in *21st International Conference on Advanced Information Networking and Applications Workshops (AINAW 2007)* (2007), pp. 421–428
37. S. Duangphasuk, M. Warasart, S. Kungpisdan, Design and accountability analysis of a secure sms-based mobile payment protocol. in *8th International Conference on Electrical Engineering/Electronics, Computer, Telecommunications and Information Technology (ECTI-CON 2011)* (2011), pp. 442–445
38. ESET Latin America's Lab, Trends for 2013: astounding growth of mobile malwaree (2013), http://go.eset.com/us/resources/white-papers/Trends_for_2013_preview.pdf
39. A.P. Felt, M. Finifter, E. Chin, S. Hanna, D. Wagner, A survey of mobile malware in the wild, in *Proceedings of the 1st ACM Workshop on Security and Privacy in Smartphones and Mobile Devices (SPSM 2011)* (2011), pp. 3–14
40. L. Francis, G. Haneke, K. Mayes, K. Markantonakis, A security framework model with communication protocol translator interface for enhancing nfc transactions, in *Sixth Advanced International Conference on Telecommunications (AICT 2010)* (2010), pp. 452–461
41. A.O. Freier, P. Karlton, P.C. Kocher, The SSL protocol version 3.0: Internet draft, March 1996, http://home.mit.bme.hu/~hornak/adatbiz/ssl3/ssl-toc.html
42. F-Secure, Worm:iphoneos/ikee.b, http://www.f-secure.com/v-descs/worm_iphoneos_ikee_b.shtml
43. Y. Fu, Q. Fu, Scheme and secure protocol of mobile payment based on RFID, in *3rd International Conference on Anti-counterfeiting, Security, and Identification in Communication (ASID'09)* (2009), pp. 631–634
44. T.S. Fun, L.Y. Beng, J. Likoh, R. Roslan, A lightweight and private mobile payment protocol by using mobile network operator, in *International Conference on Computer and Communication Engineering* (2008), pp. 162–166
45. A. Furche, G. Wrightson, Subscrip – an efficient protocol for pay-per-view payments on the internet, in *5th IEEE International Conference for Computer Communication and Networks* (1996), pp. 16–19
46. J. Gao, K. Edunuru, J. Cai, S. Shim, P2P-paid: a peer-to-peer wireless payment system, in *The Second IEEE International Workshop on Mobile Commerce and Services (WMCS 2005)* (2005), pp. 102–111
47. J. Gao, V. Kulkarni, H. Ranavat, L. Chang, H. Mei, A 2d barcode-based mobile payment system, in *Third International Conference on Multimedia and Ubiquitous Engineering* (2009), pp. 320–329
48. J. Gao, A. Küpper, Emerging technologies for mobile commerce. J. Theor. Appl. Electron. Commer. Res. **1**(1), Editorial (2006)
49. J.Z. Gao, L. Prakash, R. Jagatesan, Understanding 2d-barcode technology and applications in m-commerce – design and implementation of a 2d barcode processing solution, in *Computer Software and Applications Conference (COMPSAC'07)* (2007), pp. 49–56
50. Gartner, Inc., Gartner says annual smartphone sales surpassed sales of feature phones for the first time in 2013 (2014), http://www.gartner.com/newsroom/id/2665715
51. M. Girault, Self-certified public keys, in *EUROCRYPT* (1991), pp. 490–497
52. S. Glassman, M. Manasse, M. Abadi, P. Gauthier, The millicent protocol for inexpensive electronic commerce, in *Fourth International World Wide Web Conference* (1995), pp. 603–618
53. R.M. Godbole, A.R. Pais, Secure and efficient protocol for mobile payments, in *10th international conference on Electronic commerce (ICEC 2008)* (2008)
54. J.A.O. Gonzalez, Multi-party non-repudation protocols and applications. Ph.D. thesis, University of Malaga, Campus de Teatinos (2006)
55. J.A. Guerrero-Ibáñez, C. Flores-Cortés, S. Zeadally, Vehicular ad-hoc networks (VANETs): architecture, protocols and applications, in *Next-Generation Wireless Technologies: 4G and Beyond*, ed. by N. Chilamkurti, S. Zeadally, H. Chaouchi (Springer, London, 2013), pp. 49–70. ISBN:978-1-4471-5164-7
56. W. Guo, Design of architecture for mobile payments system, in *Control and Decision Conference (CCDC 2008)* (2008), pp. 1732–1735

57. T. Halonen, A system for secure mobile payment transactions. Master's thesis, Department of Computer Science, Helsinki University of Technology (2002)
58. W. Ham, H. Choi, Y. Xie, M. Lee, K. Kim1, Secure one-way mobile payment system keeping low computation in mobile devices, in *The Third International Workshop on Information Security Applications (WISA 2002)* (2002), pp. 287–301
59. K. Hansen, T. Larsen, K. Olsen, On the efficiency of fast RSA variants in modern mobile phones. Int. J. Comput. Sci. Inf. Secur. (IJCSIS) **6**(3), 136–140 (2009)
60. J. Hao, J. Zou, Y. Dai, A real-time payment scheme for SIP service based on hash chain, in *IEEE International Conference on e-Business Engineering (ICEBE 2008)* (2008), pp. 279–286
61. H. Harb, H. Farahat, M. Ezz, Securesmspay: secure SMS mobile payment model, in *2nd International Conference on Anti-counterfeiting, Security and Identification (ASID 2008)* (2008), pp. 11–17
62. M. Hashemi, E. Soroush, A secure m-payment protocol for mobile devices, in *Canadian Conference on Electrical and Computer Engineering (CCECE 2006)* (2006), pp. 294–297
63. R. Hauser, M. Steiner, M. Waidner, Micro-payments based on iKP, in *14th Worldwide Congress on Computer and Communications Security Protection* (1996), pp. 67–82
64. H. van der Heijden, Factors affecting the successful introduction of mobile payment systems, in *Proceedings of the 15th Bled eCommerce Conference* (2002), pp. 430–443
65. A. Herzberg, Payments and banking with mobile personal devices. Commun. ACM **46**(5), 53–58 (2003)
66. S. Hillman, C. Neustaedter, E. Oduor, C. Pang, Mobile payment systems in north america: user challenges & successes, in *Proceedings of the Extended Abstracts of the 32nd Annual ACM Conference on Human Factors in Computing Systems (CHI EA 2014)* (2014), pp. 1909–1914
67. Z.-Y. Hu, Y.-W. Liu, X. Hu, J.-H. Li, Anonymous micropayments authentication (AMA) in mobile data network, in *23rd Annual Joint Conference of the IEEE Computer and Communications Societies (INFOCOM 2004)* (2004), pp. 46–53
68. R.-J. Hwang, S.-H. Shiau, D.-F. Jan, A new mobile payment scheme for roaming services. Electron. Commer. Res. Appl. **6**(2), 184–191 (2007)
69. R.-J. Hwang, F.-F. Su, L.-S. Huang, Fast firmware implementation of RSA-like security protocol for mobile devices. Wirel. Pers. Commun. **42**(2), 213–223 (2007)
70. J.T. Isaac, J.S. Cámara, An anonymous account-based mobile payment protocol for a restricted connectivity scenario, in *18th International Workshop on Database and Expert Systems Applications (DEXA 2007)* (2007), pp. 688–692
71. J.T. Isaac, J.S. Cámara, Anonymous payment in a client centric model for digital ecosystems, in *Inaugural IEEE International Conference on Digital Ecosystems and Technologies (IEEE-DEST 2007)* (2007), pp. 422–427
72. J.T. Isaac, J.S. Camara, A.I. Manzanares, M.C. Castro, Payment in a kiosk centric model with mobile and low computational power devices, in *International Conference of Computational Science and Its Applications (ICCSA 2006)* (2006), pp. 798–807
73. J.T. Isaac, J.S. Cámara, S. Zeadally, J.T. Márquez, A secure vehicle-to-roadside communication payment protocol in vehicular ad hoc networks. Comput. Commun. **31**(10), 2478–2484 (2008)
74. J.T. Isaac, S. Zeadally, An anonymous secure payment protocol in a payment gateway centric model. Procedia CS **10**, 758–765 (2012)
75. J.T. Isaac, S. Zeadally, Design, implementation, and performance analysis of a secure payment protocol in a payment gateway centric model. Computing **96**(7), 587–611 (2013)
76. J.T. Isaac, S. Zeadally, Secure mobile payments. IT Professional **16**(3), 36–43 (2014, in press)
77. J.T. Isaac, S. Zeadally, J.S. Camara, Implementation and performance evaluation of a payment protocol for vehicular ad hoc networks. Electron. Commer. Res. **10**(2), 209–233 (2010)
78. J.T. Isaac, S. Zeadally, J.S. Camara, A lightweight secure mobile payment protocol for vehicular ad-hoc networks (vanets). Electron. Commer. Res. **12**(1), 97–123 (2012)

79. ISACA, Mobile payments: risk, security and assurance issues, Nov 2011, http://www.isaca.org/Groups/Professional-English/pci-compliance/GroupDocuments/MobilePaymentsWP.pdf, Last accessed 30 Sept 2013
80. ISO/IEC 9594-8, ITU-T X.509 recommendation. Information Technology — Open Systems Interconnection — The Directory: Authentication Framework (1993)
81. C. Jennings, J. Fischl, H. Tschofenig, G. Jun, Payment for services in session initiation protocol (SIP). draft-jennings-sipping-pay-06.txt, IETF (2007)
82. Y.-S. Jeong, N. Sun, S.-H. Lee, IPTV micropayment system based on hash chain using RFID-USB module, in *IEEE 34th Annual Computer Software and Applications Conference* (2012)
83. E.W. Kelley, Jr., The future of electronic money: a regulator's perspective. IEEE SPectrum, Special Issue on Electronic Money (1997), pp. 20–22
84. JSON, Introducing JavaScript object notation (JSON) (1999), http://json.org/
85. S. Kadhiwal, M.A.U.S. Zulfiquar, Analysis of mobile payment security measures and different standards. Comput. Fraud Secur. **2007**(6), 12–16 (2007)
86. J. Kadirire, The short message service (SMS) for schools/conferences, in *Recent Research Developments in Learning Technologies, ULTRALAB, Anglia Polytechnic University* (2005), pp. 977–981
87. P. Kieseberg, M. Leithner, M. Mulazzani, L. Munroe, S. Schrittwieser, M. Sinha, E. Weippl, QR code security, in *Proceedings of the 8th International Conference on Advances in Mobile Computing and Multimedia (MoMM 2010)* (2010), pp. 430–435
88. B. Kim, Design of fair tracing e-cash system based on blind signature. Master's thesis, School of Engineering Information and Communications University (2004)
89. K. Krombholz, P. Frühwirt, P. Kieseberg, I. Kapsalis, M. Huber, E. Weippl, QR code security: a survey of attacks and challenges for usable security, in *Proceedings of Human Aspects of Information Security, Privacy, and Trust – Second International Conference, HAS 2014, Held as Part of HCI International 2014* (2014), pp. 79–90
90. N. Kshetri, Mobile payments in emerging markets. IT Professional **14**(4), 9–13 (2012)
91. D. Kumar, Y. Ryu, A brief introduction of biometrics and fingerprint payment technology, in *Second International Conference on Future Generation Communication and Networking Symposia (FGCNS 2008)* (2008), pp. 185–192
92. S. Kungpisdan, Modelling, design, and analysis of secure mobile payment systems. Ph.D. thesis, Monash University (2005)
93. S. Kungpisdan, B. Srinivasan, P.D. Le, Lightweight mobile credit-card payment protocol, in *4th International Conference on Cryptology in India (Progress in Cryptology – INDOCRYPT'2003)* (2003), pp. 295–308
94. S. Kungpisdan, B. Srinivasan, P.D. Le, A secure account-based mobile payment protocol, in *International Conference on Information Technology: Coding and Computing (ITCC 2004)* (2004), pp. 35–39
95. S. Kungpisdan, M. Warasart, Somp: an SMS-based operator- assisted mobile payment protocol, in *The 1st International Computer Science and Engineering Conference (ICSEC 2010)* (2010), pp. 29–34
96. X. Lai, J.L. Massey, A proposal for a new block encryption standard, in *Workshop on the Theory and Application of Cryptographic Techniques on Advances in Cryptology (EUROCRYPT 1990)* (1991), pp. 389–404
97. C. Lamprecht, A. van Moorsel, P. Tomlinson, N. Thomas, Investigating the efficiency of cryptographic algorithms in online transactions. Int. J. Simul. Syst. Sci. Technol. **7**(2), 63–75 (2006)
98. V. Lazarov, N. Degefa, S. Seid, Secure mobile payments (2008), http://home.in.tum.de/~lazarov/html-data/files/research/papers/payments-report.pdf, Last accessed Dec 2012
99. J. Lee, C.-H. Cho, M.-S. Jun, Secure quick response-payment (QR-pay) system using mobile device, in *13th International Conference on Advanced Communication Technology (ICACT'11)* (2011), pp. 1424–1427
100. Y. Lei, D. Chen, Z. Jiang, Generating digital signatures on mobile devices, in *18th International Conference on Advanced Information Networking and Applications (AINA'2004)* (2004), pp. 532–535

101. X. Li, The role of mobile agents in m-commerce, in *The Sixth Wuhan International Conference on E-Business (WHICEB 2007)* (2007), pp. 403–408
102. Q. Li, G. Clark, Mobile security: a look ahead. IEEE Secur. Priv. **11**(1), 78–81 (2013)
103. V. Li, X. Hu, L. Zeng, The application of mobile agent in mobile payment, in *International Conference on Computer Science and Network Technology (ICCSNT 2011)* (2011), pp. 1612–1616
104. W. Li, Q. Wen, Q. Su, Z. Jin, An efficient and secure mobile payment protocol for restricted connectivity scenarios in vehicular ad hoc network. Comput. Commun. **35**(2), 188–195 (2012)
105. P. Limpittaya, M. Warasart, S. Kungpisdan, Design and analysis of a secure agent-based mobile bill payment protocol for bulk transactions, in *International Joint Conference on Computer Science and Software Engineering (JCSSE 2012)* (2012), pp. 71–76
106. P. Lindgren, C. Olsson, Peer-to-peer technology. Technical report, Chalmers University of Technology (2006)
107. W. Liu, C. Zhao, W. Zhong, Z. Zhou, F. Zhao, X. Li, J. Fu, K.S. Kwak, The GPRS mobile payment system based on RFID, in *International Conference on Communication Technology (ICCT 2006)* (2006), pp. 1–4
108. J.L.-C. Lo, J. Bishop, J.H.P. Eloff, Smssec: an end-to-end protocol for secure SMS. Comput. Secur. **27**(5–6), 154–167 (2008)
109. H. Lu, F. Claret-Tournier, C. Chatwin, R.C.D. Young, M-commerce secured using web-enabled mobile biometric agents, in *IEEE/WIC/ACM International Conferences on Web Intelligence and Intelligent Agent Technology – Workshops (WI-IATW 2007)* (2007), pp. 480–483
110. R.R. Maddirala, Secure file transfers using mobile device. Master's thesis, Department of Computer Engineering and Computer Science California State University, Long Beach (2010)
111. L. Mainetti, L. Patrono, R. Vergallo, IDA-pay: an innovative micro-payment system based on NFC technology for android mobile devices, in *20th International Conference on Software, Telecommunications and Computer Networks (SoftCOM 2012)* (2012)
112. M.S. Manasse, The millicent protocols for electronic commerce, in *Proceedings of the First USENIX Workshop on Electronic Commerce, USENIX* (1995)
113. R. Martínez-Peláez, F. Rico-Novella, C. Satizábal, Mobile payment protocol for micropayments: withdrawal and payment anonymous, in *2nd International Conference on New Technologies, Mobility and Security (NTMS 2008)* (2008), pp. 1–5
114. Mastercard and Visa, Set protocol specifications book 1–3 (1997)
115. K.N. McGill, Trusted mobile devices: requirements for a mobile trusted platform module. J. Hopkins Apl Tech. Dig. **32**(2), 544–554 (2013)
116. G. Me, M.A. Strangio, A. Schuster, Mobile local macropayments: security and prototyping. IEEE Pervasive Comput. **5**(4), 94–100 (2006)
117. J. Meng, L. Ye, Secure mobile payment model based on wap, in *4th International Conference on Wireless Communications, Networking and Mobile Computing (WiCOM 2008)* (2008), pp. 1–4
118. D.S. Milojicic, V. Kalogeraki, R. Lukose, K. Nagaraja, J. Pruyne, B. Richard, S. Rollins, Z. Xu, Peer-to-peer computing. Technical report, HP Laboratories Palo Alto (2003), http://www.hpl.hp.com/techreports/2002/HPL-2002-57R1.pdf
119. D.P. Mirembe, J. Kizito, D. Tuheirwe, H.N. Muyingi, A model for electronic money transfer for low resourced environments: M-cash, in *Third International Conference on Broadband Communications, Information Technology & Biomedical Applications* (2008), pp. 389–393
120. Modern Survival Blog, Foods for barter (2012), http://modernsurvivalblog.com/survival-kitchen/foods-for-barter/. Last accessed Dec 2012
121. E.A. Morse, V. Raval, PCI DSS: payment card industry data security standards in context. Comput. Law Secur. Rev. **24**(6), 540–554 (2008)
122. C.R. Mulliner, Security of smart phones. Master's thesis, University of California Santa Barbara (2006)

123. S.K. Nair, E. Zentveld, B. Crispo, A.S. Tanenbaum, Floodgate: a micropayment incentivized P2P content delivery network, in *17th International Conference on Computer Communications and Networks (ICCCN 2008)* (2008), pp. 1–7
124. S. Nakamoto, Bitcoin: a peer-to-peer electronic cash system (2008), https://bitcoin.org/bitcoin.pdf. Last accessed July 2015
125. J. Nan, L. Xiang-dong, Z. Jing-ying, Y. De-li, A mobile micropayment protocol based on chaos, in *Eighth International Conference on Mobile Business (ICMB 2009)* (2009)
126. H. Neumann, T. Schwarzpaul, Digital coins: fairness implemented by observer. J. Theor. Appl. Electron. Commer. Res. **1**(1), 1–15 (2006)
127. E.W.T. Ngai, F.K.T. Wat, A literature review and classification of electronic commerce research. Inf. Manag. **39**(5), 415–429 (2002)
128. T.N.T. Nguyen, P. Shum, E.H. Chua, Secure end-to-end mobile payment system, in *2nd International Conference on Mobile Technology, Applications and Systems* (2005)
129. M. Niranjanamurthy, N. Kavyashree, S. Jagannath, R. Bhargava, M-commerce: security challenges issues and recommended secure payment method. Int. J. Manag. IT Eng. **2**(8), 374–393 (2012)
130. NIST, Fips pub 46-3 data encryption standard (DES) (1999), http://csrc.nist.gov/publications/fips/fips46-3/fips46-3.pdf
131. NIST, Fips pub 197 advance encryption standard (AES) (2001), http://csrc.nist.gov/publications/fips/fips197/fips-197.pdf
132. Ntt DoCoMo, I-mode, http://www.nttdocomo.co.jp/
133. T. Okamoto, K. Ohta, Universal electronic cash, in *Proceedings of the 11th Annual International Cryptology Conference on Advances in Cryptology (CRYPTO 1991)* (1992), pp. 324–337
134. D. O'Mahony, M.A. Peirce, H. Tewari, *Electronic Payment Systems for E-Commerce* (Artech House, Boston, 2001)
135. E. Oswald, Introduction to elliptic curve cryptography (2005), http://www.ne.lpfritz.org/ftp/ElipticLast accessed Jan 2013
136. B. Ozdenizci, K. OK Vedat Coskun, M.N. Aydin, NFC loyal: a beneficial model to promote loyalty on smart cards of mobile devices, in *International Conference for Internet Technology and Secured Transactions (ICITST 2010)* (2010), pp. 1–6
137. J. Peha, I. Khamitov, Paycash: a secure efficient internet payment system, in *5th International Conference on Electronic Commerce (ICEC 2003)* (2003), pp. 125–130
138. M. Peirce, Multi-party electronic payments for mobile communications. Ph.D. thesis, University of Dublin, Trinity College (2000)
139. H. Petersen, P. Horster, Self-certified keys: concepts and applications, in *Third International Conference on Communications and Multimedia Security* (1997), pp. 102–116
140. H. Pieterse, M.S. Olivier, Security steps for smartphone users, in *Information Security for South Africa (ISSA 2013)* (2013), pp. 1–6
141. A. Pîrjan, D.-M. Petroşanu, A comparison of the most popular electronic micropayment systems. Roman. Econ. Bus. Rev. **3**(4), 97–110 (2008)
142. M.L. Polla, F. Martinelli, D. Sgandurra, A survey on security for mobile devices. IEEE Commun. Surv. Tutorials **15**(1), 446–471 (2013)
143. P. Pourghomi, G. Ghinea, Ecosystem scenarios for cloud-based NFC payments, in *Proceedings of the Fifth International Conference on Management of Emergent Digital EcoSystems (MEDES 2013)* (2013), pp. 113–118
144. Psion Teklogix, Psion, http://www.psionteklogix.com
145. N. Rai, A. Ashok, J. Chakraborty, P. Arolker, S. Gajera, M-wallet: an sms based payment system. *International Journal of Engineering Research and Applications – Special Issue of the National Conference On Emerging Trends in Engineering & Technology* (2012), pp. 258–263
146. E. Ramezani, Mobile payments (2008), http://webuser.hs-furtwangen.de/~heindl/ebte-08-ss-mobile-payment-Ramezani.pdf, Last accessed Dec 2012

147. F. Reid, M. Harrigan, An analysis of anonymity in the bitcoin system, in *Security and Privacy in Social Networks*, ed. by Y. Altshuler, Y. Elovici, A.B. Cremers, N. Aharony, A. Pentland (Springer, New York, 2013)
148. R. Rivest, A. Shamir, L. Adleman, A method for obtaining digital signatures and public-key cryptosystems. Commun. ACM **21**(2), 120–126 (1978)
149. R.L. Rivest, A. Shamir, Payword and micromint: two simple micropayment schemes. CryptoBytes **2**(1), 7–11 (1996)
150. A. Romão, M. Mira da Silva, An agent-based secure Internet payment system for mobile computing, in *International IFIP/GI Working Conference on Trends in Distributed Systems for Electronic Commerce (TREC 1998)* (1998), pp. 80–93
151. J. Rosenberg, H. Schulzrinne, G. Camarillo, A. Johnston, J. Peterson, R. Sparks, M. Handley, E. Schooler, SIP: session initiation protocol. RFC 3261, IETF (2002)
152. A. Ruiz-Martínez, C.I. Marín-López, A lightweight payment scheme for real-time services based on SIP. EURASIP J. Wirel. Commun. Netw. **2012**(161), 1–25 (2012)
153. N. Sadeh, *M-Commerce, Technologies, Services, and Business Models* (Wiley, New York, 2002)
154. I. Satoh, Mobile agents, in *Coordination of Large-Scale Multiagent Systems* (Springer, New York, 2006), pp. 231–254
155. G.P. Schneider, *Comercio Electronico* (I.T.P. Latin America, 2004)
156. B. Schoenmakers, Basic security of the ecash payment system, in *Applied Cryptography, Course on Computer Security and Industrial Cryptography* (1997), pp. 201–231
157. S. Schwiderski-Grosche, H. Knospe, Secure mobile commerce. Electron. Commun. Eng. J. (2002)
158. S. Schwiderski-Grosche, H. Knospe, Secure mobile commerce. Electron. Commun. Eng. J. **14**(5), 228–238 (2002)
159. V.C. Sekhar, M. Sarvabhatla, A secure account-based mobile payment protocol with public key cryptography. ACEEE Int. J. Netw. Secur. **3**(1), 5–9 (2012)
160. V.C. Sekhar, M. Sarvabhatla, Secure lightweight mobile payment protocol using symmetric key techniques, in *International Conference on Computer Communication and Informatics (ICCCI 2012)* (2012), pp. 1–6
161. S.M. Shedid, M. El-Hennawy, M. Kouta, Modified set protocol for mobile payment: an empirical analysis, in *2nd International Conference on Software Technology and Engineering (ICSTE 2010)* (2010), pp. 350–355
162. K. Siau, H. Sheng, F. Fui-Hoon Nah, The value of mobile commerce to customers, in *Third Annual Workshop on HCI Research in MIS* (2004), pp. 65–69
163. D. Singelée, B. Preneel, The wireless application protocol (WAP) (2003)
164. B. Singh, K.S. Jasmine, Comparative study on various methods and types of mobile payment system, in *International Conference on Advances in Mobile Network, Communication and its Applications (MNCAPPS 2012)* (2012), pp. 143–148
165. E.L. Slade, M.D. Williams, Y.K. Dwivedi, Mobile payment adoption: classification and review of the extant literature. Mark. Rev. **13**(2), 167–190 (2013)
166. Smart Card Alliance, The mobile payments and nfc landscape: a U.S. perspective (2011), http://www.smartcardalliance.org/resources/pdf/Mobile_Payments_White_Paper_09(1611)pdf
167. X.J. Song, Mobile payment and security, in *T-110.501 Seminar on Network Security 2001* (2001), Poster paper 8
168. P. Soni, M-payment between banks using SMS [point of view]. Proc. IEEE **98**(6), 903–905 (2010)
169. A.K. Sood, S. Zeadally, Drive-by download attacks: a comparative study of browser exploit packs features and attack techniques. IEEE IT Prof. (2016, in press)
170. B. Sun, Y. Xiao, K. Wu, Intrusion detection in cellular mobile networks, in *Wireless Network Security*, ed. by Y. Xiao, X.S. Shen, D.-Z. Du (Springer, New York, 2007), pp. 183–210
171. Sun Microsystem, Java platform, micro edition (Java me), API specification (2008), http://java.sun.com/javame/index.jsp

172. Sun Microsystem, Java platform, micro edition (Java se) v 1.6.0, API specification (2008), http://java.sun.com/javase/index.jsp
173. A.A. Tabandehjooy, N. Nazhand, A lighweight and secure protocol for mobile payments via wireless internet in m-commerce, in *International Conference on e-Education, e-Business, e-Management, and e-Learning (IC4E 2010)* (2010), pp. 495–498
174. J. Téllez, J. Sierra, A. Izquierdo, M. Carbonell, Anonymous payment in a kiosk centric model using digital signature scheme with message recovery and low computational power devices. J. Theor. Appl. Electron. Commer. Res. **1**(2), 1–11 (2006)
175. The Legion of the Bouncy Castle, The legion of the bouncy castle Java cryptography APIs version 1.4 (2008), http://www.bouncycastle.org/
176. The National Institute of Standards and Technology (NIST), Guidelines on Hardware-Rooted Security in Mobile Devices (Draft), Special Publication 800-164, Oct 2012
177. B.T.S. Toh, S. Kungpisdan, P.D. Le, Ksl protocol: design and implementation, in *IEEE Conference on Cybernetics and Intelligent Systems* (2004), pp. 544–549
178. M. Toorani, A.A.B. Shirazi, Ssms – a secure SMS messaging protocol for the m-payment systems, in *13th IEEE Symposium on Computers and Communications (ISCC 2008)* (2008), pp. 700–705
179. S. Töyssy, M. Helenius, About malicious software in smartphones. J. Comput. Virol. **2**(2), 109–119 (2006)
180. D.M. Tripathi, A. Ojha, Lpmp: an efficient lightweight protocol for mobile payment, in *3rd National Conference on Emerging Trends and Applications in Computer Science (NCETACS 2012)* (2012), pp. 41–45
181. Y-M. Tseng, J.K. Jan, H.-Y. Chien, Digital signature with message recovery using self-certified public keys and its variants. Appl. Math. Comput. **136**(2–3), 203–214 (2003)
182. T. Tsiakis, G. Sthephanides, The concept of security and trust in electronic payments. Comput. Secur. **24**(1), 10–15 (2005)
183. U. Varshney, Mobile payments. IEEE Comput. **35**(12), 120–121 (2002)
184. J. Viega, B. Michael, Guest editors' introduction: mobile device security. IEEE Secur. Priv. **8**(2), 11–12 (2010)
185. A. Vilmos, S. Karnouskos, Semops: design of a new payment service, in *14th International Workshop on Database and Expert Systems Applications (DEXA 2003)* (2003), pp. 865–869
186. S. Viveros, The economic impact of malicious code in wireless mobile networks, in *4th International Conference on (Conference Publication No. 494) 3G Mobile Communication Technologies (3G 2003)* (2003), pp. 1–6
187. X.F. Wang, K.-Y. Lam, X. Yi, Secure agent-mediated mobile payment, in *PRIMA* (1998), pp. 162–173
188. Webopedia, Windows phone, http://www.webopedia.com/TERM/W/windows_phone.html
189. M. Whaiduzzaman, M. Sookhak, A. Gani, R. Buyya, A survey on vehicular cloud computing. J. Netw. Comput. Appl. **40**, 325–344 (2014)
190. Wikipedia, Windows phone 8, http://en.wikipedia.org/wiki/Windows_Phone_8
191. X. Yi, C.K. Siew, X.F. Wang, E. Okamoto, A secure agent-based framework for internet trading in mobile computing environments. Distrib. Parallel Databases **8**(1), 85–117 (2000)
192. H.-C. Yu, K.-H. Hsi, P.-J. Kuo, Electronic payment systems: an analysis and comparison of types. Technol. Soc. **24**(3), 331–347 (2002)
193. M.J. Yuan, *Enterprise J2ME: Developing Mobile Java Applications* (Prentice Hall PTR, Upper Saddle River, 2003)
194. S. Zeadally, R. Hunt, Y.-S. Chen, A. Irwin, A. Hassan, Vehicular ad hoc networks (VANETS): status, results, and challenges. Telecommun. Syst. **50**(4), 243–346 (2012)
195. G. Zhang, F. Cheng, C. Meinel, SIMPA: a SIP-based mobile payment architecture, in *IEEE/ACIS International Conference on Computer and Information Science (ICIS 2008)* (2008), pp. 287–292
196. R. Zhang, J.Q. Chen, C.A.J. Lee, Mobile commerce and consumer privacy concerns. J. Comput. Inf. Syst. **53**(4), 31–38 (2013)

197. J. Zhang, W. Zou, D. Chen, Y. Wang, On the security of a digital signature with message recovery using self-certified public key. Informatica (Slovenia) **29**(3), 243–346 (2005)
198. X. Zhang, O. Aciiçmez, J.-P. Seifert, Building efficient integrity measurement and attestation for mobile phone platforms, in *Security and Privacy in Mobile Information and Communication Systems, First International ICST Conference, MobiSec 2009*, Turin, 3–5 June 2009, Revised selected papers, ed. by A.U. Schmidt, S. Lian. Volume 17 of Lecture Notes of the Institute for Computer Sciences, Social Informatics and Telecommunications Engineering (Springer, Berlin/Heidelberg, 2009), pp. 71–82
199. X. Zheng, D. Chen, Study of mobile payments system, in *IEEE International Conference on E-Commerce (CEC 2003)* (2003), pp. 24–27
200. Y. Zhu, A new architecture for secure two-party mobile payment transactions. Master's thesis, Faculty of Arts and Science, University of Lethbridge (2010)
201. ZippyCart, Mcommerce revenues in Europe to surpass us by end of 2010, http://www.zippycart.com/ecommerce-news/1202-mcommerce-revenues-in-europe-surpass-us-by-end-of-(2010)\discretionary-html, Last accessed Dec 2012

# Index





J. Téllez, S. Zeadally, *Mobile Payment Systems*, Computer Communications and Networks, DOI 10.1007/978-3-319-23033-7

The manufacturer's authorised representative in the EU is Springer Nature Customer Service Centre GmbH, Europaplatz 3, 69115 Heidelberg, Germany. If you have any concerns regarding our products, please contact ProductSafety@springernature.com

Printed and bound by CPI Group (UK) Ltd, Croydon, CR0 4YY
01/10/2026
02225883-0006